evolution

Morton Jenkins

TEACH YOURSELF BOOKS

Acknowledgements
The author expresses his sincere thanks to Joanne Osborn of Hodder & Stoughton for generating the idea of the book and for her help and encouragement throughout its production. It is also a pleasure to thank Dr Sue Noake, Headteacher of Lewis Girls Comprehensive School, Ystrad Mynach for her advice and help during the preparation of the text.

For UK order queries: please contact Bookpoint Ltd, 78 Milton Park, Abingdon, Oxon OX14 4TD. Telephone: (44) 01235 400414, Fax: (44) 01235 400454. Lines are open from 9.00–6.00, Monday to Saturday, with a 24-hour message answering service. Email address: orders@bookpoint.co.uk

For U.S.A. & Canada order queries: please contact NTC/Contemporary Publishing, 4255 West Touhy Avenue, Lincolnwood, Illinois 60646–1975, U.S.A. Telephone: (847) 679 5500, Fax: (847) 679 2494.

Long renowned as the authoritative source for self-guided learning – with more than 40 million copies sold worldwide – the *Teach Yourself* series includes over 200 titles in the fields of languages, crafts, hobbies, business and education.

British Library Cataloguing in Publication Data
A catalogue record for this title is available from The British Library.

Library of Congress Catalog Card Number: On file

First published in UK 1999 by Hodder Headline Plc, 338 Euston Road, London, NW1 3BH.

First published in US 2000 by NTC/Contemporary Publishing, 4255 West Touhy Avenue, Lincolnwood (Chicago), Illinois 60646–1975 U.S.A.

Typeset by Transet Limited, Coventry, England.
Printed in Great Britain for Hodder & Stoughton Educational, a division of Hodder Headline Plc, 338 Euston Road, London NW1 3BH by Cox & Wyman Ltd, Reading, Berkshire.

Impression number 10 9 8 7 6 5 4 3 2
Year 2005 2004 2003 2002 2000

CONTENTS

INTRODUCTION

> It has sometimes been said that the success of the Origin [of Species] proved that the subject was in the air, or that men's minds were prepared for it. I do not think that this is strictly true, for I occasionally sounded out a few naturalists, and never happened to come across a single one who seemed to doubt about the permanence of species … I tried once or twice to explain to able men what I meant by Natural Selection, but signally failed.
>
> Charles Darwin

Perhaps Charles Darwin was too modest, although cynics have been known to describe evolution as 'the incomprehensible taught by the incompetent'. The challenge is to condense four thousand million years into a book.

We are members of the most exclusive club in our solar system – the dominant species on the third planet from a fairly average star. As thinking human beings we are the only organisms on Earth that can be ashamed of the fact that we are animals. When we look at the living world we can see other forms of life, some commonplace and others so strange that we may gasp in amazement when we encounter them. Yet all the living things on Earth are results of adaptation to trials or benefits provided by selective pressures of the environment. They have obeyed the rule 'adapt or become extinct' and have survived due to the kinds of changes that are encompassed by the term evolution.

The concept of evolution is arguably one of the most important ideas in modern history, so it may seem strange that it is so little understood among what is undoubtedly one of the best educated and most intellectually sophisticated societies of all time. Why is the idea so readily accepted and routinely used by some while being such anathema to others? Why is such a biological foundation stone the target for so many critics' sledge hammers?

There are probably two major reasons: one is based on some people's perception that it is a theory which is at odds with religious dogma; the second is an uneasiness due to a misunderstanding of its principles. For the first type of critic, matters of faith lie outside the realm of scientific investigation. Arguments aimed by either side clearly and inevitably miss their targets. However, while aiming to provide a scientifically objective and unbiased account of evolution, this book considers the idea of creationism in its own context in the first chapter.

The major aim of the book is to clear up some misunderstandings while building a foundation that will help to clarify the meaning of evolution. In the simplest terms, evolution is change through time – or, as Charles Darwin phrased it, 'descent with modification'. It is important also to keep in mind that, in this context, *individuals* do not evolve; *populations* evolve. So the book is an attempt to present a brief account of what is known or what is postulated about the origin of life and evolutionary changes.

Although Darwin's concepts regarding evolution, published in 1859 in *The Origin of Species*, were eventually accepted by most of the scientific community of his time, critical questions arose immediately. In particular, people asked, 'just how can natural selection be driven? What are the mechanisms?' The answers lay directly in the area of genetics that the first 'mathematical biologist', Gregor Mendel, was to elegantly elucidate in 1865, after twenty years of research. Just how close did Darwin come to learning about Mendel's original research? The journal in which Mendel's work was published was found in Darwin's library; however, whereas the adjacent article was heavily marked with Darwin's annotations, Mendel's paper was left in pristine condition. Apparently Darwin was not looking into anything of a mathematical nature for an explanation of mechanisms of natural selection and simply did not grasp Mendel's conclusions. Instead, with each subsequent edition of *The Origin of Species*, Darwin retreated more and more towards explanations which hinted at inheritance of acquired characteristics, the approach that so deeply flawed the theory of Darwin's predecessor, Jean Baptiste de

Lamarck. While in school, Darwin once wrote that he was 'mired firmly in the mud of mathematics' and that is where he would remain!

After the rediscovery of Mendel's principles at the very beginnings of the twentieth century, there followed an upsurge of intrigue in Mendelian genetics. People began to see that the Mendelian principles provided an explanation of the mechanism for natural selection that had eluded Darwin and other evolutionists for so long. By the 1940s, Darwin's original theory was firmly established and integrated with Mendelian genetics, forming what is now called 'neo Darwinism'. This book focuses on neo Darwinism but takes the reader through the long – and, at times, frustrating – history of its development. However, before natural selection can take place, there must be a starting point on which selective pressures can work.

About four thousand million years ago a unique phenomenon took place which changed the face of the Earth: the emergence of life. This book attempts to address questions such as what sparked off the creation of living forms and why the evolutionary process did not stop after reaching the first types best suited to the environment. Perhaps the most talked-about ideas of evolution refer to explanations of the emergence of humans and whether there will be further progression. These ideas and an explanation of how birds and mammals managed to overtake reptiles provide material for one of the chapters together with an account of the roles played by the great geological catastrophes which have influenced the history of the Earth.

1 FACT OR FICTION?

The fact of change

Imagine if our planet was visited by aliens from other galaxies. Their observations of human behaviour during the last century might read as follows:

> They kill the largest living land animal, make balls from its teeth and knock them into holes in a table with sticks. They call it snooker.
> They take the gut of small animals, stretch it on a wooden frame and knock rubber balls backwards and forwards. They call it tennis.
> They take leaves from a plant, dry them, roll them into paper, put them in their mouths and set fire to them. They call it smoking.
> Their technology is advanced enough to carry them to the moon in space ships. They call it progress.

The very complex brain of the human is capable of creating all of the above activities. However, if aliens had visited our planet several millions of years ago, they would have reported very different facts about human activity. Perhaps their report would have been something like this:

> They are covered in hair. They kill animals with sticks and eat them raw. They sometimes walk on two legs but usually crawl on all fours with their hands touching the ground.

Was this animal the same as humans of the twenty-first century? It took about six million years for this ancestor of humans to develop or change into our present-day form. This process of gradual change is called evolution and has continued to take place throughout the history of Earth. Plants and animals are continually changing as a result of selective pressures of their environments, so that the best-adapted types survive to breed and pass on their genes. Those that fail to adapt become extinct and at best remain as fossils.

How are these changes taking place?

The history of an idea

Throughout human history new scientific ideas regularly appear. Some of these prove to be erroneous but often provide at least a framework for the development of knowledge. However, many of the embryonic ideas which have been expressed by scientists and which have progressed into basic scientific principles that we take for granted today, began with less than total peer enthusiasm. The formation of hypotheses that did not fit into current knowledge, often faced antagonism and blatant hostility. Science has not progressed smoothly up a gradient from the ancient world to the present.

Writings from the early civilizations of Babylon, Egypt and Greece show evidence that the great minds of the time debated the origins of life. Dusty parchments found in monasteries showed that the ancient scholars increasingly sought answers to thought-provoking questions in the religious authorities of the time. Some attempted to explain the nature of life in a world that was beyond the understanding of mere mortals. Consequently, science and religion soon became inextricably entangled. The age of investigation to verify predictions had not yet dawned. Statements that were not testable often became dogma and were therefore not questioned. Thus the Church became the seat of higher learning on one hand, and on the other an opponent of new ideas which went against religious doctrines.

The line between heretical thinkers and searchers of the truth was cobweb thin, as was demonstrated by the fates of several radical scholars of the sixteenth century. **Copernicus** (1473–1543) expressed his idea that the Earth was not, after all, the centre of the universe. As a consequence, he escaped the ultimate in retribution only by dying shortly after publication of his work. The evidence of **Galileo** (1564–1642) that the Earth actually orbited the Sun was considered so outrageous by the Roman Catholic Church that he was forced to publicly deny his belief before the Inquisition in Rome. He was not officially forgiven until the 1980s!

Mysticism and magic played a big part in the minds of the world's most important decision makers well into the eighteenth century but a surge of scientific thinking had begun at the time of **Isaac Newton** (1642–1727), with the formulation of laws of motion based on the ideas of gravitation. Until the 18th century, science was largely limited to disciplines involving what we now know as mathematics, astronomy and physics: the word **biology** was yet to be invented. Indeed, the study of living things was thought to be more suitable for the minds of philosophers and religious thinkers. To probe the essence of life would elicit dreadful consequences according to theologians. Churchgoers were told that life must have a special purpose and some grand design. The human race was simply not ready to consider itself as just another physical phenomenon.

Most scientists of that era also believed that all species were created in their present form – that is, they had not changed during their time on Earth. However, more and more people were observing the variety of living things. Herbalists and 'quacks' were in abundance and needed names for their products. The rudiments of the systematic study of living things goes back to **Aristotle** (384–322 BC), who made an attempt at classifying organisms into groups, the members of which had certain features in common. Although this was a crude division largely based on superficial features, Aristotle's system was logical and had some fundamental principles which can still be used today. It was the Swedish botanist, **Carl von Linne** (1707–1778), who devised a system of classification for all known organisms, naming them in Latin. This proved to be the basis of an international method of naming every living thing. Incidentally, his love of the Latin language influenced von Linne to change his own name to **Carolus Linnaeus** in his published works. Linnaeus also envisaged species as being unchangeable and thought of them as products of the divine creation. He attempted to describe species in precise terms and in doing so became aware of the difficulties created by the fine details which distinguish closely related varieties that appear via hybridization. The seeds of thoughts which challenged the immutability of species had already been sown and were beginning to germinate in the minds of certain revolutionaries.

Most of his contemporaries accepted Linnaeus' views but there was an exception, the Frenchman **George-Louis Leclerc de Buffon** (1707–1788), who proposed that, in addition to those animals that were the products of creation, there were types 'conceived by Nature and produced by time'. He explained that changes of this kind were the results of imperfections in the Creator's expression of the ideal. He suggested that the donkey had developed from the horse by a sort of degeneration, and likewise monkeys from men. In fact, he had dipped his toes in very cold water. The theologians of the day reminded him succinctly and crisply of the words of Genesis and so he removed his toes to a drier and more comfortable place in haste! If his ideas had come to the attention of the papacy a century earlier, it is likely that de Buffon would have been tried for heresy.

A decade later in 1763, **Erasmus Darwin** (1731–1802), the grandfather of arguably the most famous biologist of all time, added impetus to the idea that all species were not immutable. Considered to be a brilliant man, he earned a lucrative living as a physician and excelled as a naturalist and poet. It is possible that he acted as a catalyst for the future development of the revolutionary ideas of his grandson, Charles Darwin, because his financial success enabled Charles to have '... ample leisure from not having to earn [his] own bread' (F. Darwin, 1958). He recognized the importance of competition in the formation of species, the effects of the environment on changes in species, and the possibility of the inheritance of these changes. Charles Darwin was born eight years after his grandfather's death but, despite this gap, Charles lived much of his life in the intellectually broad shadow of Erasmus.

Others also began to think that species could change and that the changes could be inherited. In France, **Jean Baptiste Lamarck** (1744–1829) was a protégé of de Buffon. One of his claims to fame is the invention of the name biology but his major contribution to science was his work on evolution. He suggested that not only had one species arisen from another but that humans had arisen from another species. Again, here was a bold statement that would have been considered outrageous a few decades before. Lamarck believed that every organism has its position on the **scale of Nature**, with humans established at the top. He also observed that the fossils found in older layers of rock did not seem to be as complex as those in more recently deposited rock. His observations led him to conclude that older species had gradually given rise to more recent ones. In his attempt to explain this process of change, Lamarck's hypothesis was flawed because he suggested that an organism could generate new structures or organs to meet its needs. He went on to state that, once formed, such structures continue to develop through use and that their development in the parents was inherited by the offspring. The classic Lamarckian example of a scientific error was his theory of how the giraffe developed its long neck. He maintained that the long neck evolved as each generation of giraffes stretched to reach the leaves at the top of trees and that this characteristic was passed on to future generations. For this he was derided by the scientific fraternity of the day – an academic élite which was already steeped in the tradition of special creation. So Lamarck's name has come to be associated with failure and discredit, but in writing off Lamarck we write off the first serious modern attempt at a unified science of biology – and, indeed, Lamarck was a biologist *par excellence*. Many of his works remain impressive over two centuries after their publication. Perhaps his greatest contribution to the history of biology was that he put evolution 'on the map' to an extent which would in due course help the general acceptance of the ideas of Charles Darwin. He died at 85, a blind pauper, and was buried in an unmarked trench, with his daughter providing the poignant epitaph 'Posterity will remember you.'

One reason that Lamarck had little influence on his contemporaries was because his views were opposed strenuously by the most powerful and influential scientific figure in France, if not in Europe, at the time – the famous **Georges Cuvier** (1769–1832). Cuvier strongly supported the doctrine of **special creation**, to which he added the theory of **catastrophism**, which held that the Earth had been the scene of a number of violent cataclysms, each of which wiped out all life, and that new life was created following each of these upheavals. Each cataclysm buried the plants and animals of the preceding era and this accounted for the fossils of many extinct species with which Cuvier was familiar.

By now, you might be curious about the obvious 'French connection' with studies into evolution in the eighteenth century. It was at that time that French science, encouraged by public interest in natural history, was enjoying a period of great vitality. The *Jardin du Roi*, established in Paris in 1640 by Louis XIII, had grown into a centre of botanical and zoological excellence of some repute. Also, as it approached revolution, France had become the European centre of philosophy, with the result that the natural world assumed an importance not conceivable in earlier times.

The intellectual climate in Britain was more conservative than in France. Atheists were viewed as evil cranks and dismissed out of hand. It was against this background of extreme potential antagonism to the idea of evolution that the scene was set for one of the most important scientific theories ever to be put forward. *The Origin of Species by Natural Selection* shook the scientific and non-scientific world to its very core. The ownership of the theory belongs to one of the most famous scientists the world has ever known – **Charles Darwin** (1809–1882).

The great British naturalist

In the 1830s few would have regarded Charles Darwin as being destined for fame. His father was a physician of some note and had thought that his son would want to pursue a similar career. He therefore arranged a medical school education for Charles when he was 16 years old. Legend has it that he fled from the operating theatre during his first experience of the horrors of nineteenth-century surgery and never returned. Charles Darwin decided that he would never become a doctor but was persuaded to study theology. He spent three years at Cambridge University with a view to becoming an Anglican clergyman. As a member of the British upper class, Darwin spent most of his time indulging in horse riding, hunting, good food, and the occasional game of blackjack rather than in his studies. He joined the many amateur naturalists of Georgian society whose wealthy backgrounds enabled them to indulge in such hobbies. He became well liked by his many friends but did not please his father with his non-studious behaviour. His father often wondered aloud about his son, whose academic prowess had been so appallingly unremarkable that he is quoted as shouting 'You care for nothing but shooting, dogs, and rat catching, and you will be a disgrace to yourself and all your family!' What was to become of him?

It is ironic to think that but for a chance turn of fate Darwin might have become an inconspicuous pastor, tending his parish and studying botany in his spare time. The window of opportunity opened when Darwin's friend and mentor, Professor Reverend John Henslow, informed him of an offer of free passage on a survey ship, HMS *Beagle*, which was to chart foreign waters on a world-wide voyage of oceanographic discovery. In 1831, Henslow wrote to Darwin telling him that Captain Robert Fitzroy was willing 'to give up part of his own cabin to any young man who would volunteer to go with him without pay as a naturalist to the voyage of the *Beagle.'* The voyage was intended to last for five years. In some ways, the position looked unattractive – no pay and sleeping in a hammock in a crowded chartroom – but hesitation is rare in youth and Darwin was no exception. After all, he was 22 years old with the whole world in front of him – literally.

Confident, with Henslow's recommendation and his recently obtained degree in theology, Darwin eagerly applied to the equally young Captain Fitzroy for the post. Fitzroy had been placed in command of the *Beagle* in 1829 after its previous commander, Stokes, had committed suicide – a fate that destiny had in store for Fitzroy many years later. He was a strict believer in the words of the Bible and was convinced that any observations and conclusions that Darwin would make on the voyage would add to his faith in the wonders of life created by the Almighty. It is said that he almost turned down Darwin for the position because of the shape of his nose (he believed that the shape of the nose reflected the character of its bearer and that Darwin's nose was weak)!

When Darwin approached his father with the proposal, his father's reaction was, perhaps, predictable. He thought that the idea was scarcely suitable for a prospective clergyman and agreed only if Charles could find a sponsor to provide money for his keep. It so happened that one of Darwin's uncles was Josiah Wedgwood, of pottery fame and wealth, and he agreed to provide the necessary funds. Even when all the arrangements had been made and his voyage about to begin, Darwin still showed some unease, which is clear in this letter to his sister, Susan:

> Fitzroy says the stormy sea is exaggerated: that if I do not choose to remain with them, I can at any time get home to England, and that if I like, I shall be left in some healthy, safe and nice country, that I shall always have assistance; that he has many books, all instruments, guns are at my service. There is indeed a tide in the affairs of men, and I have experienced it. Dearest Susan, Goodbye.

After being driven back twice to harbour by heavy seas, on 27 December 1831 HMS *Beagle* finally weighed anchor and set out on a journey which was to be the spark for a flame that would set the scientific world alight. The retreating shoreline of Devonport blushed as the rising sun painted it, and the 25-m, 242-tonne solid barque-rigged brig creaked its way towards South America. It was then that Darwin's worst fears materialized. The romantic morning departure changed into a nauseous routine of tough shipboard life and endless seasickness. He sometimes spent whole days below deck, but fortunately Fitzroy treated Darwin's lack of sea legs with sympathy, allowing him to share his table for meals and giving up his cabin when Darwin later became afflicted with a tropical disease.

The Enchanted Isles

When Darwin set out on the voyage of the *Beagle*, he had no disagreement with the current belief that life had originated through special creation and that species were fixed for ever more. He was conscious of the fact that many scientists believed that they should be discovering how nature worked and that they should use their observations to show the wisdom of the Creator. The physical sciences were less restricted by religious dogmas and there were hints in these disciplines to show that some scientists would later find it relatively easy to fall in line with Darwin's train of thought. Among these scientists was **Charles Lyell** (1797–1875), who developed bold ideas in his *Principles of Geology*, the first volume of which was published before the *Beagle* set sail. Darwin took it on the voyage and arranged for Volume II to be sent to him while he was away. These classic works of science greatly influenced

Darwin in his thinking. Lyell rejected the biblical view that the Earth had been created in 4004 BC, proposing that it was very much older. He suggested that the Earth's changing form was a result of slow, steady processes that took place over an exceedingly long period. Darwin realized the implications of this statement. The whole concept of time was changing. This is probably the most difficult concept of all for the human mind to comprehend. Modern methods of dating rocks can be used to tell us that the true age of the Earth is of the order of four thousand million years. But what does four thousand million years really mean? Can we imagine such a length of time?

On arrival in several South American ports or landfalls, Darwin eagerly went ashore to explore. He was fascinated by the variety of animals and plants that he observed together with rich fossil beds. Early in his explorations, he was struck by how living things could vary so much from one place to the next. For example, shells from the Atlantic coast were not like those from the Pacific shoreline. He noted that in some cases species of birds and mammals changed gradually from one place to another, one type giving way to another almost unnoticeably. In other cases, one kind of organism would suddenly disappear, another having appeared in its place.

It was his visit to the **Galapagos Islands** in 1835, however, that proved to be most important when it came to proposing his theory of natural selection. The Galapagos or 'Enchanted Isles' comprise a chain of islands, 580 miles (about 900 km) off the coast of Equador. The islands became the cradle of an idea that history would most closely associate with Darwin. Captain Fitzroy was concerned with charting the relatively unknown waters and harbours to prepare for potential trade with Britain, although very few true oceanographic observations were made from the *Beagle*. Fitzroy made extensive observations of tides in the Pacific while Darwin collected many biological samples in shallow waters and developed his ideas on the origin and development of coral reefs. On the morning of 17 September, they anchored off San Cristobal, one of the more rugged and barren islands, and soon Darwin was on his way with his collecting gear and note books. His enthusiasm continued while landing on as many of the islands as he could possibly visit.

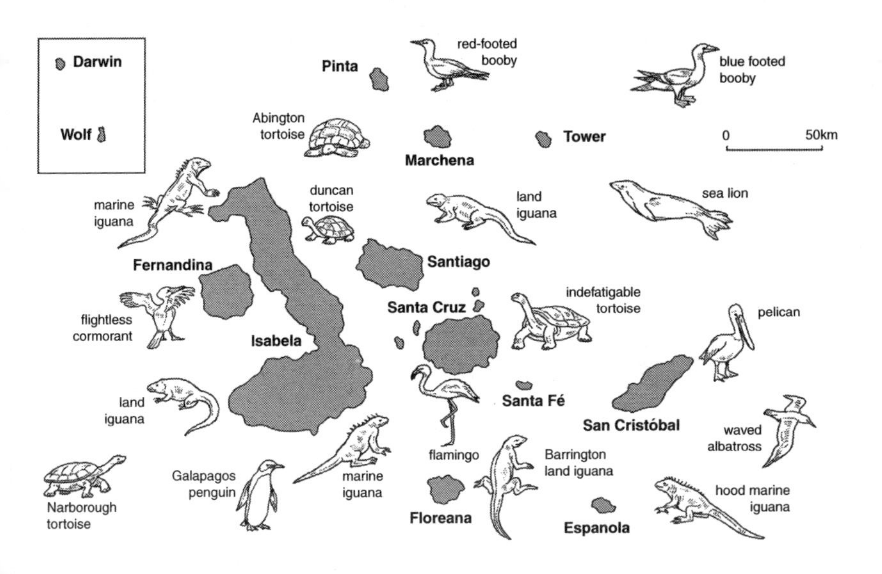

Figure 1.1 Darwin's visit to the Galapagos Islands

'A little world in itself,' Darwin wrote, 'with inhabitants such as are found nowhere else.' There were marine iguanas, half a metre long, grazing on seaweed beneath the sea. Darwin called them 'imps of darkness, black as the porous rocks over which they crawl.' He saw the giant tortoises that mariners had been used to catching and storing, live, upside-down on their decks for fresh meat on long voyages. His island visits allowed him to collect not only marine iguanas but also their land-bound relatives. Birds were abundant, many of which he was convinced were undescribed species. Among these was a drab-looking group of **finches** that were similar to those he had collected on the South American mainland. Later, Darwin was to regret not separating his collected birds according to the islands on which they were found. He noticed that two of the finches, taken from different islands, differed in the shape and size of their beaks.

The importance of these observations was emphasized by a chance conversation with Mr Lawson, the English vice-governor of the settlement of political outcasts on Charles Island. Darwin was informed that Lawson could tell from which particular island any of the giant tortoises came because they differed in the size and shape of their shells. These giant tortoises no longer live on the mainland,

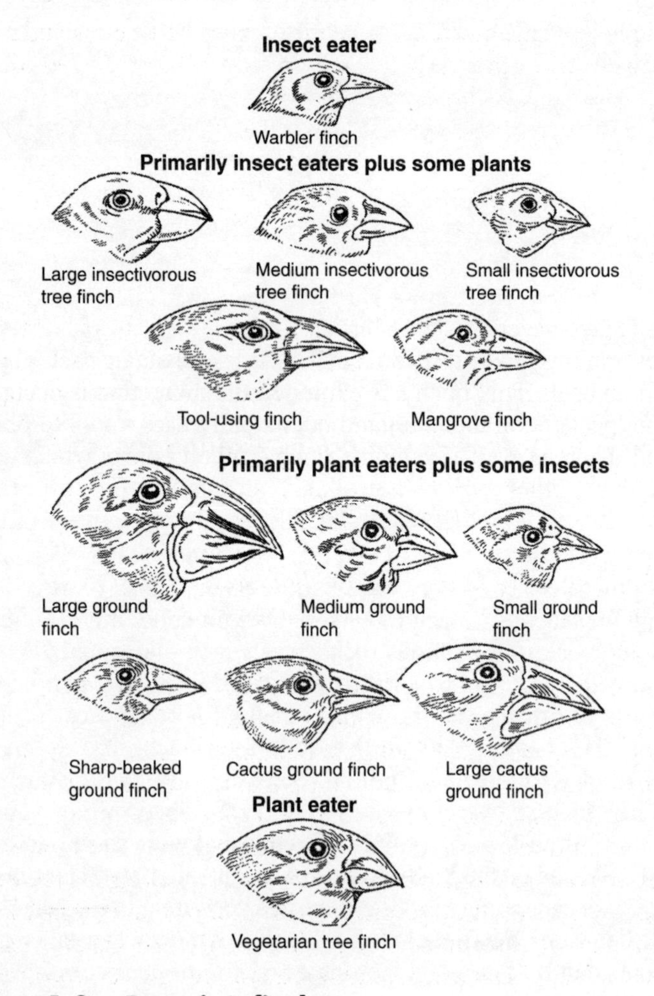

Figure 1.2 Darwin's finches

although their fossils exist there. On the Galapagos Islands today there are ten distinct races of tortoise, each from a particular location or island. Originally there were at least fifteen races but over the past 400 years five have become extinct through human activities. There is only one survivor of a particular race from the island of Pinta. 'Lonesome George' is now permanently 'home alone' at the Charles Darwin Research Station on Santa Cruz.

Having been made aware of these differences between island races of tortoise, from that day on Darwin carefully separated all his collections from each island. This was to prove to be a valuable decision once he was home in England. Years later, Darwin had his finch collection examined by a British specialist and it was decided that there were thirteen species, differing mainly in diet and shape and size of their beaks. Since then, a fourteenth species has been described.

Each species has a different-shaped beak, adapted to a particular feeding habit – seeds, insects, leaves, or parasitic ticks and lice. One species winkles out grubs from cavities in trees using a cactus spine held in its beak. This finch has exploited the niche that is occupied by woodpeckers on the mainland but actually uses a tool to obtain its food. There is even a 'vampire' finch on Wolf Island, which takes blood from other birds. It probably started on this gruesome evolutionary road by first removing ticks or other parasites from the feathers of other species of birds. A stab from a sharp beak could easily puncture the skin and provide a nutritious drink of blood. From then on it is easy to understand how these finches got hooked on the same habit as Dracula!

Darwin suggested that these birds must have come originally from the South American mainland since the volcanic Galapagos Islands would have been formed later than the mainland. In fact the oldest island, Espanola, is less than four million years old and the youngest, Fernandina, is less than 700 000 years: mere 'spring chickens' in geological terms. It is thought that the 'ancestor' finches arrived on San Cristobal and gradually colonized the other islands. Why were these so different from those on the mainland, and why did the assortment on each island differ so much from the one on the next? Darwin would not have the answer until many years later.

The voyage continued for another year and he collected vast amounts of information and specimens. On arrival home on 2 October 1836, Darwin was greeted enthusiastically by both his family and the scientific community. They wanted to know what he had seen and what he had brought back. A new phase of his work had begun.

Darwin was happy to be back among friends and familiar surroundings. He eventually married Emma Wedgwood, his first cousin. They lived in London for a while but soon moved to Down House in a small community in Kent. His wealthy background and his judicious financial investments provided him with an annual income of £1339, which was enough to support him and his family in some comfort in the 1880s. The house is now a museum devoted to Darwin. He worked earnestly on his manuscripts and during his research came across an essay by the Reverend **Thomas Malthus** (1766–1834), which was probably the first published warning of the dangers of overpopulation. Malthus's essay of 1798 pointed out that populations tended to increase in geometric progression and that if humans continued to reproduce at the same rate they would inevitably outstrip their food supply and create 'a world full of misery and vice'. Darwin applied this theological idea to his own observation that species have a high reproductive potential but that not all individuals reproduce because of differences in their survival abilities. Populations are kept in check partly because not all animals or plants survive long enough to reproduce. Malthus's essay provoked Darwin to thinking.

He calculated that even one pair of elephants, which are notoriously slow breeders, could produce 19 million progeny in only 750 years. However, over that sort of time, the population size of elephants stayed roughly the same. So what was interfering with their reproductive capacity? Were some less successful at producing offspring than others? If this were the case, Darwin wondered what factors determined which ones were to be successful.

Darwin was familiar with the principle of **artificial selection**, although the intricacies of the genetic explanation were not revealed until the work of Gregor Mendel was recognized at the beginning of the twentieth century. For thousands of years it had been known that through careful selection of animals and plants for breeding, certain desired traits could be passed on from one generation to the next. For example, by mating only the offspring of the best milk producers, breeders have developed the high-yielding dairy herds with which we are familiar today. By carefully choosing only the offspring of the best-laying hens for breeding, our battery hen 'egg factories' have been developed. Perhaps the most obvious example is seen in the vast number of varieties of dogs, which are all of the same species but have all been artificially selected from a wolf-like ancestor. Darwin demonstrated how artificial selection could be applied to pigeons and he bred many varieties at his home in Kent. In fact, this was probably the only experimental evidence that he had to reinforce his theory of **natural selection**. He deduced that nature could take the place of humans in the selection process and could determine the reproductive success of individuals. Darwin suggested that natural selection would be far more 'hit-or-miss' than artificial selection because individuals with only somewhat less desirable characteristics might still be able to produce at least some offspring and so their traits would take longer to disappear from the population. On the other hand, individuals with traits that were positively to their disadvantage would leave no offspring so those traits would more quickly disappear from the population. If a trait gave a reproductive advantage to a population, it could be expected to increase in frequency through the generations.

Darwin's legacy

In essence, natural selection means the process through which some types of organisms are more successful at breeding than others of the same species and so are more successful in passing on the characteristics that led to their success. Today, our knowledge of genetics has confirmed this idea but Darwin was many years ahead of the thoughts of the classic geneticists and molecular biologists of the twentieth century.

As the concept of natural selection depends on inequalities among populations, there must be a cause for the inequalities. What were the sources of these variations among offspring? Darwin proposed that variations appear at random. He recognized that there was no need for a divine driving force to manipulate change in a certain direction. Given sufficient time, a new variation could spread through a population – provided that it was of a selective advantage. In contrast to Lamarck's misconception of the explanation of the evolution of the giraffe's neck, Darwin's explanation would be as follows:

- If the long neck of the giraffe is inherited and is helpful in obtaining food, giraffes with longer necks will be better nourished and stronger. It is likely that these would have more reproductive energy and would leave more offspring.
- Among the offspring some would have longer necks than others, and these in turn, will be more successful than their shorter necked siblings.
- Thus, long-necked giraffes would be favoured through generations, resulting in a tendency for any generation to be composed of animals with longer necks than those of the preceding generation.
- Of course, there must be a limit to the length to which a giraffe's neck can grow but this would be limited by other factors such as the ability for the heart to pump blood to the brain.

It is very important to understand that Darwin developed his ideas without hard proof. There was no experimental evidence he could offer, and to make things more difficult he knew almost nothing of genetics.

On the European continent, a monk was searching for an explanation of inheritance. His name was **Gregor Mendel** (1822–1884) and, in 1865, while Darwin was still puzzling over the enigma of heredity, wondering how characteristics are actually transmitted, Mendel presented his theory to the Brunn Natural History Society. This was the birth of **genetics** as we know it, although its significance was not recognized until 1901. Darwin's ignorance of genetics probably delayed the publication of his theory of evolution because he felt that he should be able to explain the mechanism by which variation appeared.

In 1858 a twist of fate acted as a catalyst for the publication of one of the most important scientific books of all time. Darwin received an unfinished research paper from a young biologist, **Alfred Russel**

Wallace (1823–1913), who was working in Indonesia. Wallace sought Darwin's opinion of the merits of his work, although he had arrived at many of Darwin's conclusions independently. In scientific research, being first to publish often means being foremost in the particular field and so Darwin was almost startled into publishing his theory by this correspondence from Wallace. At the same time, encouragement to publish came from many eminent scientists, including the geologist Lyell, the botanist Joseph Hooker, and the scientific philosopher Thomas Huxley. Darwin requested that Wallace should be allowed to present his research paper to the Linnean Society of London, a society of world-wide repute, the membership to which all notable biologists aspire. The request was turned down and the two research papers were read at the same meeting, Darwin's first. The papers were published in August 1858 and Darwin's *The Origin of Species* became available to the public a year later. The first edition was sold out on the first day of publication!

Opinions of the book ranged from those of members of the public, theologians, and philosophers who thought it heretical, through those of some scientists who dismissed it because it lacked experimental evidence, to those who greeted it with enthusiasm. A conflict emerged between the concrete ideas of certain theologians and those prepared to consider an alternative to the special creation theory. It was a battle that Darwin could not fight alone because of ill health, which afflicted him on his return from the voyage of the *Beagle*. His illness was at one time thought to be due to an infection called Chagas' disease, induced by being bitten by a benchuga bug in South America, but this idea has now been largely discounted. Consequently Darwin became withdrawn from the great debate over natural selection that he had instigated. However, some of the most brilliant logical thinkers of the time came forward to defend his ideas whenever debate was necessary. The sheer logic of Darwin's theory put forward a unifying concept which accounted for his observations. He used three facts and two deductions based on them:

- The first fact was stated in Malthus's *Essays on Population*: **organisms have a tendency to multiply in geometric progression** – i.e. 2, 4, 8, 16, 32 and so on.

- The second fact was that **the numbers of a species tend to remain constant over long periods of time**.
- The first deduction was based on these two facts: **a struggle for existence takes place**. Organisms are in a constant competition for a chance to reproduce.
- The third fact is that **all living things vary**.
- The second deduction was natural selection: **some individuals are more likely to succeed in reproducing than others**. Those with favourable characteristics will be more likely to survive and reproduce than those with unfavourable characteristics.

Darwin's theory was not to be rejected on any basis other than a better explanation. There was none.

Scattered throughout Darwin's letters are one or two references to pain or uneasiness felt in the region of his heart but it is fairly certain that he had no serious or permanent trouble of this nature until shortly before his death. Despite the general improvement in his health in his later years, eventually there was a certain loss of physical vigour. Darwin died after a heart attack at about four o'clock on Wednesday 19 April 1882 in the seventy-fourth year of his age. On 26 April he was buried in Westminster Abbey, a few feet away from Sir Isaac Newton. The choir sang an anthem especially composed for the occasion, with words from the Book of Proverbs: 'Happy is the man who finds wisdom and getteth understanding'.

The most incisive of Darwin's biographers have portrayed him as a chronic invalid, wracked by doubt, incapable of aesthetic or religious emotion; and often as an anti-Christ of his day who ruined the faith of millions and sowed the seeds of Marxism and fascism. This is a fallacy: Darwin *did* suffer psychological problems, but the symptoms were not progressive. More importantly, he elucidated the processes which have changed biology from a simple description of a static world into a study of a dynamic and exciting series of interactions between organisms and the environment. Darwin found wisdom and died a happy man. The following sentences were added to his Autobiography in 1879:

> As for myself, I believe that I have acted rightly in steadily following and devoting my life to Science. I feel no remorse

> from having committed any great sin, but have often and often regretted that I have not done more direct good to my fellow creatures.

The great controversy

> And out of the ground the Lord God formed every beast of the field, and every fowl of the air; and brought them unto Adam to see what he would call them; and whatsoever Adam called every living creature, that was the name thereof.

If Adam reviewed the whole animal kingdom on this occasion, he must have had a tiring day, for the known species number something like 1.4 million.

Linnaeus attempted it over two hundred years ago but he used Latin to avoid the kind of ambiguity and confusion to which Adam's system is liable. He also put the animals into an orderly arrangement which powerfully suggests their natural kinship, although Linnaeus himself was no believer in evolution. This is illustrated by his own proud motto '*Deus creavit, Linnaeus disposuit*' – God created, Linnaeus arranged.

Faith and logic do not often mix and it is really pointless to try to apply science to the faith of all those with an unshakeable belief in the written words of the Bible or any other mainstay of their religion. By definition a free society deserves the right to believe whatever it pleases and to worship whatever God it believes in. The facts that virgin birth could not have given Jesus Christ a Y chromosome and that males and females actually have the same number of ribs, even though Eve was made from one of Adam's, just help to provide an endless list of controversies between scientists and theologians. Darwin stoked up the fire of controversy and even though it dies down occasionally there are enough glowing embers to rekindle it from time to time.

In April 1925, a teacher's decision to describe evolution in a few sentences to an enquiring student sparked off a legal trial that resulted in major battles between Darwinists and fundamentalists. The teacher was John Thomas Scopes, the place was Rhea County, Tennessee, and the time was one month after the governor of the state had signed a new law forbidding the teaching of evolution in the state schools. Scopes was indicted, charged with having unlawfully and willingly taught a theory that humans had descended from a lower order of animals, thus denying the story of divine creation. The trial, of course, was of Scopes, not of evolution, but it attracted enormous media coverage and caused a gathering of storm clouds over the whole county. William Jennings Bryan, retired presidential candidate and a superlative speaker, came from Florida to aid the prosecution by the anti-evolutionists. The famous liberals Clarence Darrow and Dudley Field Malone joined the defence. Throughout the heat of a particularly stifling July the trial droned on. The case of Scopes versus Genesis was argued fully and brilliantly by men who could hardly have done it better before the whole of the American public. Scopes had, on the testimony of a couple of schoolboys who managed to recall a few vestiges of what he had told them, broken the law of the state. He was sentenced to pay a fine of $100, thus becoming yet another casualty of the great on-going war between the Bible and biology. Things were improving, though – in mediaeval Europe he might have been hung, drawn and quartered for such heresy!

Many may still feel that there are major doubts about evolution and Darwinism. As recently as the 1980s the media covered a dispute, when on 7 January 1982 *The Times* carried the headline 'Darwin cleared: official'. The context of this unusually emotive first leader in this particular newspaper was the case brought by the American Civil Liberties Union in a court in Little Rock, Arkansas. The officials of the Union put forward the case that a recently enacted law requiring equal weight in state schools for the teaching of the biblical account of creation and the theory of evolution was unconstitutional. The decision was on the ground that **creation-science** is religion rather than science, and therefore forbidden under the US Constitution. The Arkansas authorities did not appeal against the decision. Perhaps relatively few people will feel that a law court is the correct place to decide on the moral implications of science, but the questions that led to the Arkansas trial are still fodder for the media.

Indeed, the distinguished scientist Sir Andrew Huxley felt need to devote most of his Presidential Address to the Royal Society in 1981 to the exposition of the inadequacy of many current criticisms of Darwinism.

The contemporary scientific debates about evolution are not about whether evolution has occurred, but about whether the Darwinian mechanism of adaptation through natural selection is a sufficient cause for change. The astronomer, Fred Hoyle, in 1981 argued that the idea that life was put together by random shuffling of constituent molecules is 'as ridiculous and improbable as the proposition that a tornado blowing through a junk yard may assemble a Boeing 747'. This probability was put forward as 10 to the power of 40 000, for the chance that the 2000 enzyme molecules will be formed simultaneously from the 20 component amino acids on a single specified occasion. In 1982, Huxley pointed out:

> … the relevant thing is the chance of some far simpler self-replicating system, capable of development by natural selection, being formed at any place on the Earth's surface, at any time within a period of 10 to the power of 8 years; the expectation of such events is wildly uncertain since we know neither the nature of the hypothetical self-replicating system nor the composition of the 'primeval soup', but it is not obviously less than unity.

One of the most important developments in evolutionary theory at present concerns the idea of **punctuated equilibria**, first put forward by N. Eldredge and S.J. Gould in 1972. These scientists pointed out that many new species begin suddenly in the fossil record, while other fossil lines remain unchanged for millions of years. However, the suddenness in geological time is very different from suddenness in genetic time. Research by A.R. Templeton in the late 1970s into speciation via genetic studies provided evidence that the splitting of a lineage can take place in tens rather than thousands of generations and by a relatively small number of gene differences. Darwin himself was quite clear that evolutionary rates are non-constant. He is explicit in *The Origin of Species* that

> … the periods during which species have been undergoing modification, though very long as measured by years, have probably been short in comparison with the periods during which these same species remained without undergoing any change, and I do not suppose that the process (speciation) … goes on continuously; it is far more probable that each form

remains for long periods unaltered, and then again undergoes modification.

Thus, there is no basic difference between punctuated equilibria and Darwinian evolution.

Other modern grounds of debate do not bear significantly on evolutionary theory. Recent developments in molecular biology and gene technology have led to analysis of DNA via genetic profiling. The findings of this type of research have strengthened Darwinian premises by showing relationships between and within species.

The greatest amount of anti-Darwinian propaganda comes from creationists rather than scientists. Although there have always been those who have doubted evolution on religious grounds, a new cult of creation-science has developed. The Evolution Protest Movement was founded as long ago as 1932 in Britain, with the stated aims of publishing scientific information supporting the Bible and demonstrating that the theory of evolution is not in accordance with scientific fact. Although it numbered some eminent scientists of the day, most of the literature was written by non-scientists. The criterion for assessing fact tended to be Scripture rather than Nature. The membership grew from 200 in 1966 to 850 by 1970, catalysed by the formation in the USA of the Creation Research Society in 1963, followed by the setting up of the Institute for Creation Research in San Diego, California, in 1970. A further development in Britain in 1972 was the Newton Scientific Association.

Creationists vary in the way they interpret the creation accounts in Genesis. Some maintain that creation was completed in six 24-hour periods, and that all the living races of mankind are descended from Noah. Others are prepared to accept that the *days* of creation in Genesis can be taken as geological eras, as long as God is accepted as the sole causative agent. Non-Christians tend to regard such attitudes as intellectually regressive, and indicative of the irrelevance of Christianity. Others are both convinced Christians and orthodox evolutionists. The views can be summarized thus:

- Christians believe that their God is active through natural as well as supernatural acts. God is the creator and at the same time is accepted as working through mutation, selection etc.

- Mankind is described in the Bible as emerging from a process similar to that of the other organisms, but then *changed into God's image*. There is no reason to believe that the process of *spiritual* creation would alter us as a species physically or physiologically, although Christians believe that it makes us distinct from the rest of living things and gives a special relationship to God through a soul. As a biological species, we are genetically related to the apes via a common ancestor.
- Mankind *made in God's image* can fall from this state. This removes one of the Christian's main problems about evolution: the belief that acceptance of evolution automatically implies that humans are still on the way up, and hence improving.

Sadly, the creationists' position in debates on evolution is that their rationale depends on evolution being an impossibility, and hence pre-empts rational discussion. In order to be fair to the creationists, a biologist may believe that God does not exist, but cannot prove it. A denial is as much an act of faith as the religious person's claim that God exists.

Having established the fact of change and the principle of natural selection, the next step is to consider how it all started. After all, natural selection has to have something to work on in the first place. So where did it all begin?

In the beginning

Ever since humanity first evolved a brain capable of questioning, perhaps the most challenging question of all has been 'How did life begin?' Even the most knowledgeable thinkers seem to feel uncomfortable with this question, and tend to react with furrowed brows and displacement responses like clearing their throats or wiping their spectacles. Lengthy philosophical discussions invariably conclude with the simple reality that no-one really is certain about how life began – or indeed, what life *is*. Biologists can list some of the characteristics associated with living things but how it all began is still a thought-provoking mystery.

Ideas about the origin of life can be traced as far back as the ancient Greeks of the sixth century BC. Perhaps the more notable of these early philosophers were **Thales** and his pupil **Anaximander**. The former supposedly lived until his nineties and was one of the 'Seven Sages' of the ancient world. He may be regarded as the founder of Greek philosophy, seeking a physical explanation of the origin of the world instead of looking to mythology for an answer. Thales maintained that water is the origin of all things and that the fundamental element of all matter is water. Anaximander attempted to explain **organic evolution**, believing that living creatures arose from water as it was evaporated by the sun. It was his belief that mankind was 'like a fish in the beginning'.

The tranquil Aegean sea provided the ancient philosophers with the ideal setting for contemplation of the mystery of life, whereas those more used to their countries being ravaged by periodic floods and geological disturbances looked at the same problem with a different approach – those such as the Babylonians (and the Jews whom they captured) arrived at a different theory called **catastrophism**.

The biblical stories of the Creation and Noah's flood are regarded today as allegorical by most Christians, Jews and Moslems. However, the Old Testament account of the Creation as a sudden cataclysmic event was accepted literally by the early Christians, and was maintained as the prevailing belief in civilized Europe until the mid 1800s. Perhaps it was an easy solution to an insoluble problem but some looked further for an explanation.

Until the last part of the nineteenth century, those who sought the truth generally believed that living things originated in one of three ways:

- They could result from the reproduction of other living creatures of their own kind.
- They could result from the reproduction of very different kinds of living creatures.
- They could form spontaneously from non-living materials.

We now believe that only the first method is logical but the history of the development of the other two hypotheses is an interesting diversion because it illustrates that some were prepared to question the biblical special creation theory even when there was a danger of being persecuted for heresy.

Like produces like

The idea of like producing like is familiar to us all and is, indeed, part of the basic biological principle of reproduction. However, another idea – that like produces unlike – was regarded for centuries as a method of reproduction.

A classic example of seventeenth-century evolutionary misconception is the explanation of the origin of the barnacle goose *Branta leucopsis*. In 1657 an interested observer linked three completely different living things by suggesting that one developed from another. Today, we know it to be nonsense, but let us consider the situation thus described.

> If you throw wood into the sea, in time worms breed in it, and these gradually grow a head, feet, wings, and lastly, feathers. When they are fully grown, they are as large geese and they fly upward as other birds do – using their wings to carry them through the air.

This observer had seen the wooden hull of a ship riddled with the ship 'worm', *Teredo navalis* (actually not a worm but a burrowing mollusc). Also on the wooden hull were specimens of the stalked goose barnacle, *Lepas anatifera*. The latter has a stalk that looks remarkably like a miniature version of a barnacle goose's neck and also shell plates that look like the colouring of the plumage of the goose. The assumption was that the ship worm changed into the goose barnacle, which, in turn, changed into a goose.

It is, of course, no more than a curious but charming myth. The point of relating this tale is to illustrate that, to the people of the 1600s, anything seemed possible in nature. Even though people must have observed domestic geese hatching from eggs, many thought that nature was using different methods to arrive at similar ends.

As biologists became increasingly familiar with marine creatures, they observed that the hulls of ships became colonized by many species of plants and animals. These living things were no more produced by the wood or the sea water than animals and plants living on land are produced by the earth. Slowly, the true explanation emerged. The 'worms' are not really worms at all, but marine molluscs related to mussels and clams. The stalked goose barnacles are crustaceans related to crabs and more 'normal' looking barnacles. It so happens that they have a vague resemblance to the shape of a bird. The myth still exits in the names 'goose barnacle' and ' barnacle goose'.

As the centuries passed, biologists began to reject entirely the hypothesis that one species of animal or plant can suddenly change into another. This method of reproduction was eventually eliminated from the list of possible explanations of life's origins.

Let there be life

On the other hand, the hypothesis of **spontaneous generation** was accepted by some biologists right up to the end of the nineteenth century. The origins of such a belief extend back to the classical times of the ancient Greek philosophers. **Aristotle** (384–322 BC) is among the greatest of all biologists and no other writer has ever had greater influence on science. A profound and original thinker, he speculated on the nature of life and believed in spontaneous generation. Statements in his *Historia Animalium* showed that he mistook populations of small fish (mullet) and eels in muddy ponds as being derived from the mud itself. The authority of his work was undisputed and his errors were perpetuated. For several centuries, Aristotle's beliefs were accepted as fact – so, perhaps, it is not surprising that the idea of spontaneous generation was widespread among educated people. To doubt the Aristotelian doctrine was to defy the evidence of the processes of reason and, what was much worse, to challenge the constituted religious authorities of the time. It was rare for scholars to observe and experiment for themselves until much later in history.

Nearly 2000 years after Aristotle taught his theories at the Lyceum in Athens, biologists were expressing similar views. In 1652, Jean-Baptiste van Helmont published a book in which he stated that if wheat grains and a dirty shirt were put in a pot, mice would be formed from the interaction of the wheat grains and the dirt in the shirt. This seems strange today but other, familiar, observations were often explained in the same way. The appearance of maggots in meat left to rot was thought to be a common example of spontaneous generation. Where else could the maggots have come from?

At least one biologist questioned these archaic explanations. He was **Francesco Redi** (1626–1697). A distinguished scholar, poet and physician to the Florentine Medici family, Redi is probably best known for his investigations into the origins of living things. He tried to demonstrate the simple hypothesis that maggots found in decaying flesh were not the product of putrefaction but were produced by natural means by flies which laid their eggs in the meat. By comparing the results of leaving fresh meat to rot in sealed and unsealed containers, he was able to conclude that 'if living causes are excluded, no living things arise'.

Redi's method of investigation was an early form of **control experiment**, now almost standard practice in biological investigations. The technique takes account of the number of variables in a problem and examines the results of testing one variable while keeping the others unchanged. The number of controls used will depend on the number of possible conditions which have to be considered. Redi's sealed and open containers were an insurance that his experiment with the sealed flask was not affected by some internal condition which would have brought about a similar result in an open flask at that particular time.

Despite the logic of Redi's conclusion, he was not a wholehearted opponent of the idea of spontaneous generation. He was puzzled by the origin of insect larvae found in many plant galls found on leaves and stems. Today we know that a female gall wasp injects an egg into a leaf or stem of a plant. The egg develops into a maggot-like young stage and the plant responds by forming a large growth, the gall, around the developing insect. Some galls may be the size of a walnut. The insect grows into an adult gall wasp and bores its way out of the gall. Redi was unaware of this life cycle: he never saw a female wasp lay its eggs in the plant and, since the gall had no visible hole in it, he concluded that the wasp formed from within the plant by spontaneous generation. It remained for one of his students, **Antonio Vallisnieri** (1661–1730), using the techniques of his teacher, to explain the natural origin of these insect larvae and thus demonstrate the wider application of Redi's statement concerning the origin of maggots.

Biogenesis versus abiogenesis

The hypothesis that living things come only from other living things is called **biogenesis**. In contrast, spontaneous generation is called **abiogenesis**.

Redi's experiments disproved the latter theory for the time being but the question arose again towards the end of the seventeenth century. The re-emergence was largely due to the invention of the microscope and observations made by **Antony van Leeuwenhoek** (1632–1723). Here was a gifted amateur, with sufficient income and leisure to indulge his curiosity. Such people, amongst whom may be considered Charles Darwin, contributed greatly to the advance of science. Their contributions were due their outstanding powers of observation together with their brilliance in logical deduction. Sadly, increasing specialization, the need for the elaborate facilities of modern laboratories, and long and formal training have eliminated amateur scientists from the pursuit of discovery in most branches of science.

Born in Delft, van Leeuwenhoek was the son of a basket worker and received only a modest elementary education. Eventually he opened a draper's shop but his real passion was the magnifying glass (the compound microscope had been developed but remained unsatisfactory until the development of the achromatic lens in the nineteenth century). Van Leeuwenhoek concentrated on the improvement of the single-lens or simple microscope and observed a wide range of materials.

A whole microcosm of life was described by this amateur scientist. His descriptions of the organisms he saw were so accurate that it is possible for the reader today to use them identify the species. Some were represented by drawings, which are even more accurate than his written descriptions. Here was a whole new world of tiny animals and plants that no one had seen before the invention of the microscope. What was the origin of these tiny creatures? Van Leeuwenhoek's opinion was clear. He used his observations as evidence that living creatures are not produced by putrefaction. However, the 'specialists' of the day were not convinced. Gradually the idea that larger life forms like mice, worms, maggots, geese and flies could be produced spontaneously was abandoned but they were not so sure about microscopic organisms and many were content to explain their appearance by spontaneous generation. There were several observations that could be interpreted in such a way. If chopped hay or some seeds were placed in pure rainwater, soon there would be a myriad of microscopic animals. Were these generated from the materials put into the water? The question could not be answered by van Leeuwenhoek's (or anyone else's) 'opinion'. Science had entered the beginning of an era where evidence was essential before a theory could be postulated. For the next three hundred years the abiogenesis versus biogenesis debate continued.

It was a fortunate coincidence that the debate was based on the same experimental material – hay infusion. Hay was readily available, easy to prepare and, with the help of increasingly more sophisticated microscopes, easy to observe. Chopped hay was boiled for 10 minutes in water and left exposed to air. For the first few days, the liquid was clear and was observed with a microscope to be free from micro-organisms. After a few more days, however, the liquid became cloudy and teemed with life. These observations were made by people of both schools of thought – the advocates of both biogenesis and abiogenesis. Both groups agreed that boiling the hay at the start would kill any living thing that might be present. Therefore, they concluded that the living creatures must have developed after the hay had cooled.

The abiogenesists explained that the living things were generated from the hay and water. To thesc, the biogenesists' explanation seemed incredible. They suggested that the air contained spores which contaminated the liquid after it had cooled. The spores then changed into active creatures. Pure air must be a complex mixture and every time people took a breath, they inhaled masses of strange creatures! So the burden of proof was with the biogenesists.

One asset that the biogenesists had was the fact that their hypothesis could be tested. Their prediction was that the microbes arose solely from airborne spores. They deduced that, if they prevented air from coming in contact with the boiled, cooled hay, microbes would not appear. As long ago as 1711, **Louis Joblot** (1645–1723) tested the hypothesis.

He boiled some hay in water for 30 minutes and separated the mixture equally into two vessels of the same volume. Before the mixture had cooled, he closed one of the vessels with parchment. The other vessel was left open. Both vessels were then left for several days. Only the vessel that remained open developed microbes. If the microbes has been spontaneously produced, they should have appeared in both vessels. Because they did not, the observations could be explained only on the basis of spores reaching the contents of the flasks through the air. The open vessel was the control.

Joblot continued to investigate further. He removed the parchment cover and found that the contents of the flask soon became colonized by microbes. In a sense, Joblot had confirmed van Leeuwenhoek's idea but the difference was that the latter's view was not supported by evidence, whereas Joblot's hypothesis was supported by observations of a well planned experiment.

Although Joblot's findings may seem convincing, the question of spontaneous generation was far from closed. The English priest and biologist **John Turbeville Needham** (1713–1781) boiled mutton gravy, poured it into a glass vial, corked the vial, and left it for a few days. It swarmed with microbes. So too did mixtures of boiled seeds. These and other such observations allowed the abiogenesists to hang on to their ideas for another hundred years.

It was the Italian abbot, **Lazaro Spallanzani** (1729–1799), who revealed the source of error in Needham's experiments. Spallanzani is considered as one of the world's first great experimental physiologists but his investigations extended into physics and mineralogy. He entered the controversy of spontaneous generation when he read Needham's reports in 1748. Not being convinced by Needham's demonstrations, he supposed that Needham had not boiled the gravy for a long enough time. He repeated the experiments, varying the length of boiling time and hermetically sealing his containers. Though the procedures were straightforward, they were disciplined and controlled, and the results could be precisely interpreted. He found that different types of microbes could withstand different boiling periods but that adequate boiling would eventually kill all of them. The infusions, thus treated, could be kept indefinitely without colonization by microbes but when restored to contact with air, the contents of the flask soon contained microbes. Needham had also sealed his containers with corks, which were not airtight.

Was this the final nail in the lid of spontaneous generation's coffin? Apparently not. Needham and his supporters contended that Spallanzani's boiling had destroyed the 'vital principle' in the air in the flasks, spoiling it for the purposes of generation.

The culmination of the long debate came in the last half of the nineteenth century. The respected and skilful French scientist **Felix-Archimede Pouchet** (1800–1872) prepared his own experimental atmosphere by adding nitrogen and oxygen to form artificial air above the boiled and cooled hay infusion. In a few days microbes appeared in the mixture. In a second experiment, Pouchet even made the water for his mixture. He burned hydrogen gas in air and then boiled this water with hay. Microbes again appeared after a few days. Pouchet concluded that spontaneous generation was possible.

It was left to one of the greatest biologists of all time to challenge Pouchet's conclusions and finally lay the idea of spontaneous generation to rest. **Louis Pasteur** (1822–1895) was Dean of the Faculty of Science at Lille in 1854, where he studied alcoholic fermentation. He suggested that Pouchet's experiment was basically flawed and set out to prove that mixtures of yeast and sugar would not be colonized by microbes when air was prevented from reaching them. Pasteur concluded that pure air without microbial spores could never contaminate sterile infusions. To test his hypothesis he took his infusions of yeast and sugar, in sealed glass flasks, high into the French Alps where he believed the air to be pure. There he broke the seals of the flasks, exposed the contents to the air and resealed them by melting the glass and fusing the flasks' necks. Of twenty flasks exposed at one place, in only one did microbes grow. Some of the flasks that Pasteur used in his experiments of 1860 are still on exhibit at the Pasteur Institute and are still sterile today.

Pouchet's observations were in fact highly accurate, but neither the materials nor the methods of his experiments were the same as those used by Pasteur. Whereas Pasteur employed an easily sterilizable liquid extract of yeast, Pouchet employed an infusion of hay, the bacteria in which were not killed by the temperature to which Pasteur subjected his flasks; this explains why the former remained clear in the presence of sterile air while the latter developed microbes.

Pasteur had demonstrated that:

- All observations and experiments thought to be examples of spontaneous generation were shown to be false.
- In no experiment could it be shown convincingly that spontaneous generation could occur.

Biogenesis is now taken for granted by biologists – but the debate lasted 300 years.

The bare necessities of life

Debating the question 'How did life arise on Earth?' is a fruitless but safe intellectual exercise – you can never be proved right or wrong! However, many very reputable scientists have spent the best parts of their careers in pursuit of the answer. Darwin himself expressed some ideas. In a letter dated 1871, he wondered if the kind of complex chemicals that make up living creatures might not have been formed somewhere in a 'warm little pond' where all the ingredients might be present. Perhaps the most difficult thing to imagine, after trying to grasp the time scale, is a lifeless Earth in the beginning. The atmosphere was totally different from that of today. The air that you are breathing at this moment will be composed of roughly 78% nitrogen, 21% oxygen, 0.03% carbon dioxide, and traces of other gases such as argon and helium. The amount of water vapour that you inhale will depend on whether you are reading this book in a steamy bathroom, a dry desert, or somewhere between these two extremes in terms of humidity. Of course, today we have the problem of measuring other, more worrying, polluting gases in our atmosphere, the amounts of which will depend on the purity of the air around you.

In 1936, the Russian biochemist **Alexander Ivanovich Oparin** published a book called *The Origin of Life*. Oparin was one of the first to study the problem of life's origin in great detail and proposed that the atmosphere of early Earth contained less oxygen and a considerable amount of hydrogen. Hydrogen is so reactive that it tends to link to many other elements, particularly nitrogen, oxygen and carbon. Hydrogen and nitrogen form ammonia; hydrogen and oxygen form water; hydrogen and carbon form methane. So, because of this basic chemistry, it is assumed that the first atmosphere was a mixture of ammonia, water and methane.

The Earth, the third planet from the Sun in our solar system, was a lifeless sphere of matter covered with a thin layer of swirling gases. At first its surface was molten and volcanic but in time it began to cool and solidify. No life was there to be startled by the piercing creaks and rumbles of the groaning, contracting and expanding rocks. Dense murky clouds reflected the fiery volcanic eruptions and thunderous lightning storms pierced the sky continuously for millions of years. Condensed water vapour filled the oceans. It would seem most inhospitable as the birthplace of life.

Most authorities think that life first appeared on Earth about four thousand million years ago. The earliest known fossilized remains were found in Western Australia in 1980. They have been dated at 3.5 thousand million years old and consist of the remains of at least five different types of living things, elongated and strand-like, resembling some bacteria that we know today. The fossilized remains are called **stromatolites** and indicate that 3.5 thousand million years ago life was already here.

The molecules that form living things today are certainly much more complex than the kind that existed when the Earth was first formed by masses of gases and dust coming together. We have to assume that the types of molecules associated with living things were formed by simpler molecules joining together: however, this assumption is flawed unless we can imagine certain changes occurring on the cooling Earth. After all, why should molecules bother to join together in the first place? The types envisaged to have existed in the early atmosphere can coexist side by side without ever joining. So what made Earth more conducive to the interaction of chemicals? The answer was probably the appearance of water and a source of energy from the Sun.

As the Earth cooled, condensation allowed water to fall from the skies continuously for millions of years. Again, the time span is almost impossible to comprehend! As the waters filled the deepest valleys and canyons of the planet, the oceans were born and became the geological collection boxes of all known minerals as they were washed into them from newly formed rocks. In these watery surroundings molecules of all types were brought together – colliding and, eventually, interacting. Their interactions would be possible only if there was a source of energy. In fact there are a number of possible sources of energy. The Sun produces a variety of types of radiation besides heat. The dynamic landscape of the Earth was constantly subjected to extremes of temperature due to volcanic activity and lightning – the provision of energy was not a problem. So, we can assume the presence of water, ammonia, methane, hydrogen, an abundance of energy and all the time in the world! Given these basic ingredients, could a primordial living 'soup' be made in the biggest kitchen ever known?

In 1953 a part of the answer to this question was suggested. It came about in the laboratories of **Harold Clayton Urey** at the University of Chicago. Urey suggested to one of his PhD students, **Stanley Lloyd Miller**, that he should set up an experiment in which energy would be supplied to a mixture of methane, ammonia, water vapour and hydrogen. Miller set up the mixture in a large glass vessel. In another glass vessel, he boiled water. The steam that was formed passed up a tube and into the gas mixture. The mixture was pushed by the steam through another tube, back into the boiling water. The second tube was kept cool so that the steam turned back into water before dripping back into the hot water. In this way the mixture was kept circulating through the system, driven by the boiling water. Miller made very certain that there were no microbes in the water to make any organic molecules and kept everything completely sterile.

Next, energy had to be supplied. Urey and Miller reasoned that two likely sources of energy were ultra-violet light from the Sun and electric discharge from lightning. However, ultra-violet light is easily absorbed by glass so there was a problem in getting enough energy through the glass into his mixture. To solve the problem Miller simulated lightning by setting up a continuous electric spark in his apparatus for a week. He was able to produce hydrogen cyanide and two types of **amino acid** from inorganic molecules.

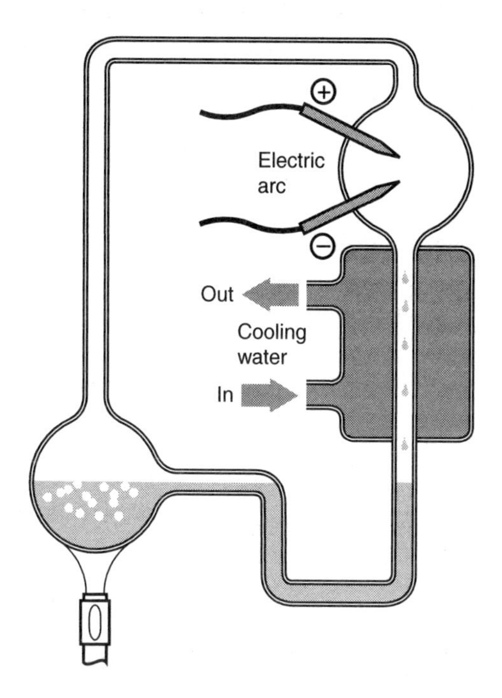

Figure 1.3 Miller's apparatus for simulating the origins of life on Earth

These were glycine and alanine, two of the twenty amino acid building blocks that go into making **protein** molecules.

With this single experiment, Miller had accomplished a great deal. His experiment had produced results within a week. What would have happened with millions of tonnes of the same chemicals, baking under the Sun's ultra-violet radiation or taking massive bolts of lightning over a thousand million years? This is the recipe for a lot of warm primordial soup!

Miller's experiment was only the beginning. Other scientists repeated it and obtained similar results. By 1968, every known amino acid had been formed in such experiments – but, of course, these were a long way away from anything resembling life. Giant molecules like **nucleotides** were essential for building **DNA**

molecules, which are universal in all living things from viruses to humans.

During their research into life's origins, it occurred to some scientists that they should begin with hydrogen cyanide, which Miller had synthesized in his primaeval atmosphere. At the University of Houston, a Spanish-born biochemist, **Juan Oro**, carried out an experiment in 1961 in which he added hydrogen cyanide to his starting mixture. He succeeded in producing amino acids that were joined together in short chains in a similar way in which they link to form proteins. Even more interesting was the fact that **purines** were formed. These are the basic building blocks of DNA. A particular purine called **adenine** was obtained – an essential component of many chemical compounds associated with living systems.

With the dawn of the 1960s, simulations of the conditions found on early Earth were producing the beginnings of nucleic acids but, besides purines, nucleic acids also require similar, but simpler compounds called **pyrimidines**. These were also synthesized in the same way. Then there were the sugars, **ribose** and **deoxyribose**, together with the **phosphate** group. The sugars proved particularly easy. Sugar molecules are made up of carbon, hydrogen and oxygen: no nitrogen is needed. Early experiments like those of Miller had produced molecules of formaldehyde (CH_2O) from carbon dioxide and water. In 1962 Oro found that if he began with formaldehyde in water and subjected the mixture to ultra-violet radiation, a variety of sugar molecules were formed. Among these were ribose and deoxyribose.

By the early 1960s then, purines, pyrimidines, ribose, and deoxyribose had all been synthesized from inorganic molecules. Phosphate groups did not have to be made. They existed in solution in the oceans. The Sri Lankan born **Cyril Ponnamperuma**, at Ames Research Centre, Moffett Field, California, conducted experiments in 1963 in which he exposed a solution of adenine and ribose to ultra-violet radiation. They linked together in just the fashion they were linked in nucleotides. When phosphate was made available, a complete nucleotide could be formed. By 1965, he and co-workers **Ruth Mariner** and **Carl Sagan** announced that they

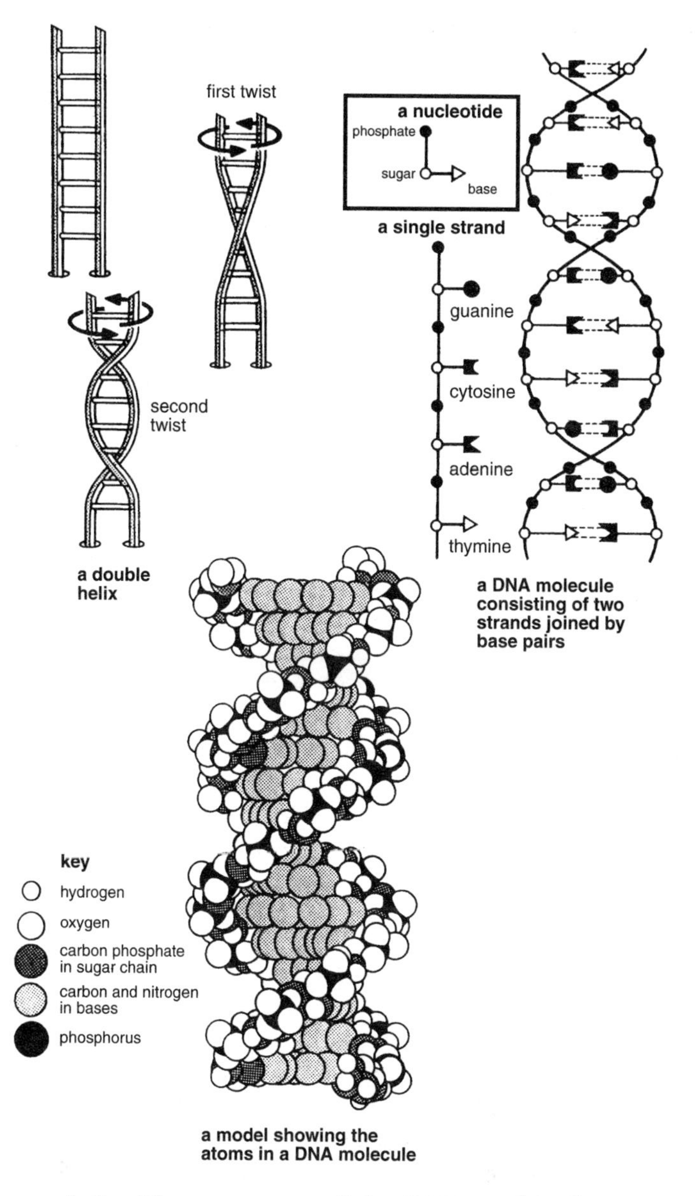

Figure 1.4 The structure of the DNA molecules

had formed a double nucleotide combined in the same way as they are in nucleic acids.

All the barc necessities of life were there on the early Earth, but life is more than just a collection of chemicals. There are thousands of delicately controlled chemical reactions going on in even the simplest living organism and these will not take place unless energy is supplied. The external supply of energy to the chemical mixtures of early Earth can be explained in physical terms but the internal energy required for activities in living things must be supplied chemically.

There are certain chemicals in living organisms which, when broken, will release energy. The energy then becomes available for living processes. Without such chemical energy, life as we know it would be impossible no matter how many proteins or nucleic acids existed. The best-known energy-rich compound is one called **adenosine triphosphate** or **ATP**. It is similar to a nucleotide to which two additional phosphate groups have been added. Ponnamperuma synthesized ATP from adenine, ribose and phosphate by prolonged irradiation with ultra-violet light. ATP is sometimes called the 'energy currency' of the cell because the energy released when it breaks down can be used for any process taking place in the cell.

The next major steps were taken by **Sidney Fox** of the University of Miami. He began his work on the origin of protein molecules in 1958 and subsequently showed how complex molecules called **polypeptides** can routinely form from mixtures of amino acids. Essentially, Fox showed that simple kinds of molecules, believed to have existed on early Earth, could be readily joined to form more complex molecules. He dripped dilute solutions of the simpler molecules onto hot sand, thereby vaporizing the water and concentrating the molecules on the substrate. The result was the formation of complex molecules he called **proteinoids**. Fox imagined the torrential rains that lasted for hundreds of millions of years splashing dilute solutions of simple molecules onto the hot rocks, where the molecules might have joined, forming complex molecules. These in turn could be washed into the oceans where they would interact.

Professor Oparin made another contribution to the explanation of how complex molecules gave rise to living organisms. He pointed out that **colloidal protein molecules**, in which the particles are suspended in a gel-like state, tend to clump together in increasingly complex masses. These may be held together by electrostatic (positive and negative) forces and thus form **coacervate** droplets. Each droplet consists of an inner cluster of colloidal molecules surrounded by a shell of water. The molecules of this water shell are arranged in a specific manner in relation to the colloid centre. As a result, there is a clear separation between the colloidal protein mass and the water in which it is suspended. Such coacervate droplets may have led to the first **protobionts** (*proto* = 'first'; *biont* = 'life form'). These are specialized droplets with internal chemical characteristics that differ distinctly from those of the external surroundings.

It is the particular orientation of the water molecules around the colloidal mass that causes the water shell to act as a barrier or **membrane**. The shell allows some molecules to pass through while preventing others and the molecules of the central colloid arrange themselves in an orderly manner depending on intermolecular attractions such as those due to electrostatic charges.

The **selectively permeable** nature of the droplets is due to the internal molecular structure. Colloids are selective in the molecules they will absorb. As a result, the internal structure of the droplet constantly changes, all the while becoming more different from its chemical surroundings. At the same time, the droplet can grow because of its continuing absorption. As this progenitor of a **cell** continues to change, the molecules at the surface of the colloid arrange themselves into a membrane-like structure just under the layer of water molecules. The membrane is even more selective than the water layer. It allows even fewer kinds of molecules to pass. Therefore, with increasing selectivity in the kinds of molecules that the mass will accept, the structure tends to become more regularly organized. Even experienced microbiologists have on occasion mistaken such droplets for microbes!

Sidney Fox has suggested an even more rapid route to the internal organization of the developing cell. He has shown that when proteinoids are mixed with cool water, the proteinoids will self-assemble into small droplets that he named microspheres. The **microspheres** can grow by absorbing proteinoids until they become so large and unstable that they fall apart, each fragment forming a new 'daughter' with constituents similar to those of the 'parent'. Another type of droplet forms when molecules called **phospholipids** are placed in water. These readily form membranes, similar to those in living cells and can break apart to form **liposomes** (*lipo* = 'fat'; *soma* = 'body'). The formation and development of the first protobionts was dependent on coacervates, microspheres and liposomes.

And then there were two

It takes another conceptual leap to imagine how protobionts could reproduce. The majority probably did not reproduce but since early Earth probably contained a vast variety of droplets, some would be more stable than others and these had the best chance of reproducing. During fragmentation, the mass of a droplet would determine its stability. Fragmentation is the very simplest form of reproduction, each fragment retaining its integrity.

A more complicated type of reproduction could be possible if great numbers of complex molecules are subjected to external changes over a sufficiently long time. Large complex molecules within the earliest protobionts were highly varied. Some would have existed as chains of simpler molecules. These chains could have attracted other molecules floating free in the fluid interior. The composition of the new chain developing along the first chain's length would have been directed by the makeup and sequence of the first chain. Suppose the second chain breaks away from the first. It then could begin to attract certain molecules from the droplet's interior and thus begin the formation of a third chain along its length. Because each segment along any such complex molecule can attract only certain other kinds of molecules, the third chain would, of necessity, be very similar to the first. This mechanism would ensure the descendency of the molecular makeup of certain successful droplets.

A successful droplet would be one which could produce the closest approximation of itself since, by its survival, it has demonstrated success.

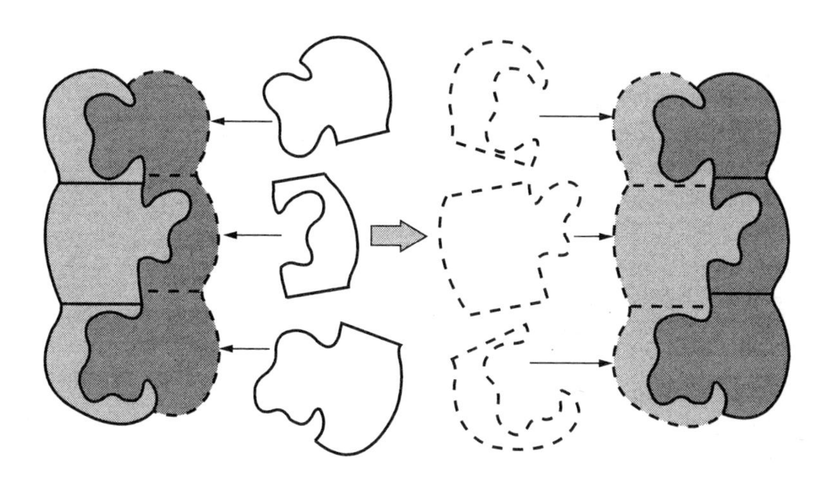

Figure 1.5 The molecule on the left serves as a template against which the molecule on the right is formed. It has been suggested that such interaction helped to ensure the continuance of the first complex molecules associated with life

The less successful droplets disappeared and the survivors continued to change, becoming ever more refined until they developed into the first cell-like bodies. Thus the ratio of the more stable forms in the population of developing cells would have continued to increase. In addition, the components of the lost and decomposing droplets would have been released and made available to the increasingly more efficient early 'cells'. All the while, the more efficient types would continue to accumulate and finally, given sufficient time the first real cells would appear.

In the absence of any solid evidence, biologists have found it easy to speculate about the nature of the first molecules needed for life.

All agree, though, that early life, by definition, must have been capable of replicating and changing through evolution. To achieve these two criteria, most biologists have assumed that the first life forms needed self-assembly instruction manuals in the form of **genes** that could be handed down from one generation to the next. Genes are made from DNA (see page 39), which is made up of nucleotides. It is the sequence of these subunits that forms a code in the instruction manual. The DNA is translated into **RNA (ribonucleic acid)**, which controls all protein production in cells. The proteins, in turn, are the labourers and skilled workers in the cellular factory. Some are enzymes which speed up the chemical reactions needed for energy conversion. Others form structural parts of the factory or are exported for useful work outside the cell. They are also used in translating DNA into RNA, and to make copies of DNA to pass to daughter cells after cell division. Take away RNA from this trio of DNA, RNA, and protein, and life grinds to a halt. Coming up with an explanation of how the trio came into existence is tricky and no one really pretends to know all the answers.

Breakthroughs sometimes appear – like suddenly finding key pieces in jigsaw puzzles – but then someone comes along and reshuffles the pieces when new hypotheses are put forward. Miller was probably the first to make a serious attempt at the jigsaw in the 1950s and since then there have been many additions. In the 1980s, **Tom Cech** at the University of Colorado and **Sydney Altman** at Yale University contributed to current thinking when they discovered that two naturally occurring RNA molecules catalyzed a reaction that snipped out regions of their own nucleotide sequence. These **catalytic RNAs** became known as **ribozymes**. There were those who leapt on this discovery, imagining an ancient world in which RNA ruled the planet. By virtue of its ability to act as a template for new RNA molecules, RNA was ideal for storing and passing on information. It also had the ability to snap bonds between atoms and could therefore be a catalyst. Crucial to the theory's credibility was the proposal that RNA once catalyzed the creation of fresh RNA molecules from their nucleotide building blocks.

Eventually RNA molecules would have acquired membranes and taken on additional catalytic jobs needed to run a primitive cellular factory. Under selective pressures, the proteins (which were more efficient catalysts than RNA) and DNA (which is less susceptible to chemical degradation) took over, relegating RNA to its present role in protein synthesis. There were more than a few sceptics, who argued that it was too great a leap from two RNA molecules undergoing a bit of self-mutilation in a test tube to managing the whole cellular factory and thereby initiating cellular organization as we know it.

In the 1990s **Jack Szostak**, a biochemist at Massachusetts General Hospital in Boston, tried to prove the sceptics wrong. He reasoned that the first prebiotic RNA molecules were assembled randomly from nucleotides dissolved in rock pools. Among trillions of short RNA molecules, there would have been one or two that could copy themselves – an ability that soon made them the dominant molecules on the planet. In order to simulate this under laboratory conditions, Szostak and his colleagues took between 100 and 1000 trillion different RNA molecules, each around 200 nucleotide units long and tested their ability to perform one of the simplest catalytic tasks – breaking another RNA molecule. They then carried out the laboratory equivalent of natural selection, selecting the few successful molecules and making millions of copies using protein enzymes. Then they mutated those RNA molecules, tested them again, replicated them again, and so on to evolve some perfect RNA-snipping ribozymes. Towards the end of the 1990s, even more efficient ribozymes were synthesized and a quote from Szostak points clearly to the future:

> We've got ribozymes doing the right kind of chemistry to copy long molecules. We haven't achieved self-replication from single nucleotides yet, but it is definitely within sight.

Some scientists have considered alternative explanations. There are those who contend that protobionts did not give rise to complex reproducing molecules. They say it is more likely that complex molecules first appeared and then built something like a cell around themselves. Others, of course, believe that life was created by some greater power and there are those who believe that life's building blocks came to Earth from outer space. In 1996, **Jeffrey Bada**, a geochemist at Scripps Institute of Oceanography found evidence that so-called Buckyballs, football-shaped molecules made of carbon atoms, had been delivered intact to Earth from outside the

Solar System. Bada and his compatriot **Luann Becker** made their discovery at Sudbury, Ontario, where a meteoroid about the size of Mount Everest crashed two thousand million years ago. This single impact site contained about one million tonnes of extraterrestrial Buckyballs. They reasoned that, if complex Buckyballs could fall to Earth without being burned up, so could complex organic molecules.

2 THE SAPLING OF LIFE

Solitary confinement

There was a vast evolutionary leap between what were essentially non-living microspheres or protobionts and cells as we know them today. Perhaps the best way to understand the levels of cellular organization that exist is to consider cells in relation to the complexity of the variety of life on present-day Earth. We are looking at the products of thousands of millions of years of evolution – although, as we shall see, evolution is continuing all around us as a never-ending phenomenon.

The rudiments of a system of classifying the variety of living things goes back to the classical scholars of the ancient world, but it was Carolus Linnaeus (see page 6) who was the first to succeed in bringing a standard format to naming organisms, therefore becoming the 'father' of the science of taxonomy. As more and more species are discovered so **taxonomy** has to accommodate them in already known groups or in newly formed groups. Therefore taxonomy itself is always evolving to allow for changing ideas concerning the affinities of living things. Today, less value is placed on classifying organisms based on their external or internal visible features and more emphasis is placed on **genetic profiling**, which indicates how close organisms are in terms of evolution. Most biologists recognize five major kingdoms. These are:

- Animalia (complex animals)
- Plantae (complex plants)
- Fungi
- Protoctista (single-celled organisms)
- Monera (bacteria and blue–green algae)

The five-kingdom classification is a starting point but there is a fundamental difference in cell structure which separates the Monera from the other four groups and is most important in understanding the evolution of cells.

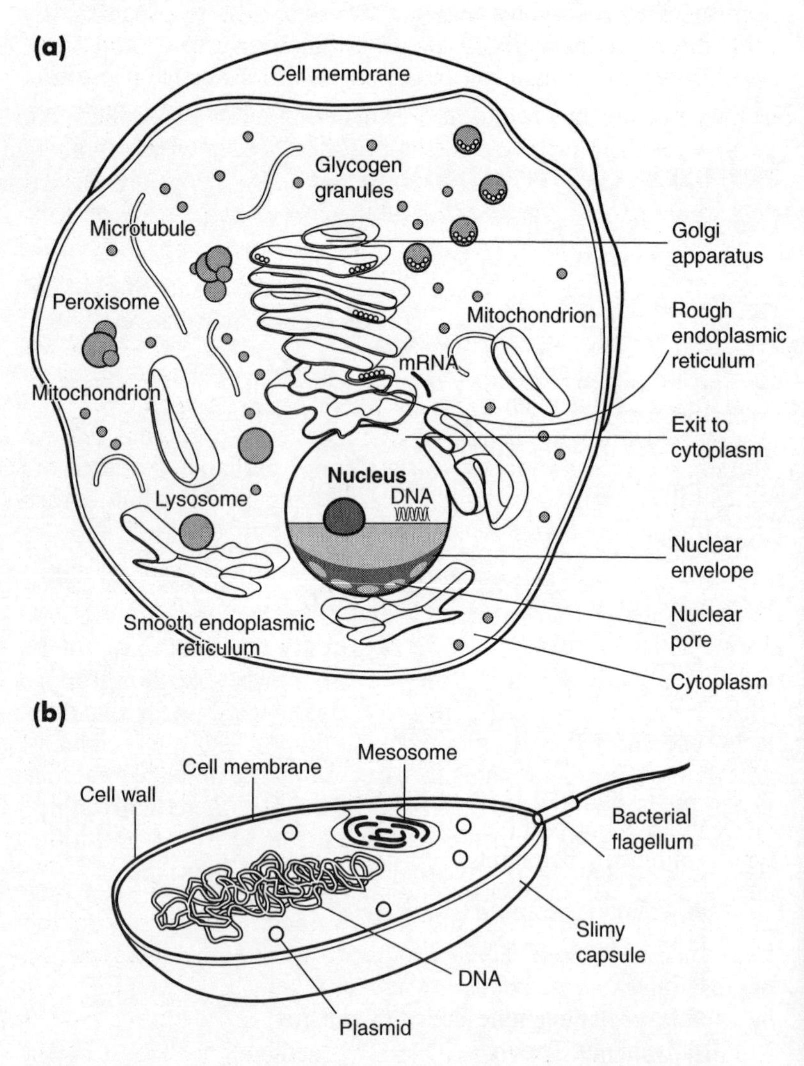

Figure 2.1 Eukaryotic (a) and prokaryotic (b) cells

The Monera are **prokaryotes** (*pro* = 'before'; *karyote* = 'nucleus') and as such lack the more sophisticated structures of the cells of the other four groups. These more complex cells are the **eukaryotes** (*eu* = 'true'; *karyote* = 'nucleus'). In every way prokaryotes are more primitive than eukaryotes and their ancestors must be considered as progenitors of the cells of modern-day forms. How did these simplest types of organisms arise from their protobiont ancestors?

Eukaryotes are not found as fossils in deposits older than 700 million years, although it is estimated that they first evolved about 1500 million years ago. As we have seen (page 35), the earliest prokaryote fossils have been found in rocks dating back to at least 3.5 thousand million years. Early prokaryotes probably consisted only of short chains of nucleotide-like units capable of replication. As in protobionts, pairs of chains could well have been joined to form more stable ring structures. Informed guesses as to the nature of ancestral prokaryotes are based on the structures found in modern types but even the most primitive cells need energy to survive and reproduce. Some early single-celled organisms like the Monera evolved a biochemical method – perhaps similar to the series of reactions involving ATP (see page 40). The blue–green algae trapped the energy of sunlight using **photosynthesis**.

The absence of oxygen in the atmosphere would have necessitated a form of **anaerobic respiration** (release of energy without oxygen), something like fermentation but using the wealth of organic chemicals that were available. After an enormous expanse of time some primitive prokaryotes might have developed the facility to join the carbon and oxygen from carbon dioxide with hydrogen from water to form simple sugars. For this to be possible, however, a chemical resembling **chlorophyll** would be needed. Cyril Ponnamperuma (see page 38) succeeded in synthesizing a type of molecule belonging to the class called **porphyrins** as long ago as 1967. He used basic chemicals which would have been available in the early stages of Earth's formation. Once chlorophyll-like molecules had been produced by adding magnesium, photosynthesis as we know it would have been possible. Oxygen, a by-product of photosynthesis, would put the purely anaerobic types at a disadvantage. They would be confined to oxygen-free habitats in mud and the abyssal zones of the oceans. Here, their raw materials for energy release are compounds of iron and sulphur. Some present-day bacteria thrive in such conditions.

If the mechanisms of energy release involved respiration, energy conversion via photosynthesis could be adopted by other cells and somehow co-ordinated and controlled. Consequently, the way would be open for the evolution of more complex cells. Such cells would be better suited to cope with the selective pressures of the ever-changing environment, allowing them to specialize and so lead eventually to the evolution of many-celled organisms.

The parts of cells

Reference to Figure 2.1 shows the fundamental differences between modern prokaryotes and eukaryotes. The structures within the cells are called **organelles**, the chief one of which is the **nucleus**. It contains all the chemical information the cell needs to function, in particular controlling the production of enzymes which in turn are responsible for regulating all the chemical reactions taking place in the whole cell. It has been suggested that some organelles evolved by the invasion of primitive cells by prokaryotes. For example, **mitochondria** (the organelles where most energy is released during respiration) are thought to be of bacterial origin, whereas the **chloroplasts** of plant cells may have evolved from invasion by blue–green algae.

Many scientists believe that the division between the prokaryotes and the eukaryotes represents the greatest single break in evolution because the differences in their cellular structures are so great. The prokaryotes have no nuclear membranes enclosing the genetic material. Instead, a single circular **chromosome** lies free in the cytoplasm. One fact that supports the idea of invasion of early cells by the ancestors of bacteria and blue–green algae is that both mitochondria and chloroplasts are similar in structure to prokaryotes. For example, they both have their own DNA, which codes for some of their proteins and much of their biochemistry is carried out on infoldings of their inner membranes.

The theory of invasion of early cells by ancestors of prokaryotes is called the **endosymbiotic theory** (*endo* = 'inside'; *symbiosis* = two different species living together for mutual benefit). It may partly account for the evolution of the eukaryotic cell but there are other reasons for this type of cell's success. Just as important for their survival was their ability to exploit and expand their capabilities in a way that the prokaryotes were unable to do. It is because of their superiority in this respect that today the eukaryotes dominate the biosphere.

Eukaryotic cells are like factories, with the nucleus representing the management, the other organelles the machinery, and the enzymes the workers. Like a factory, the cell must obtain raw materials for it to perform efficiently. It must also be able to get rid of waste. In a cell, this exchange is controlled by the **plasma membrane (cell membrane)**, which regulates the entry and exit of materials like a factory's security gates. The evolution of this membrane from its precursor in protobionts was a key step in the development of living cells. In all cells, the cell membrane is composed largely of lipid (fat) molecules called phospholipids together with protein molecules. The formation of these biochemicals – lipids and proteins – was therefore a prerequisite for the development of membranes in cells (see Figure 2.2).

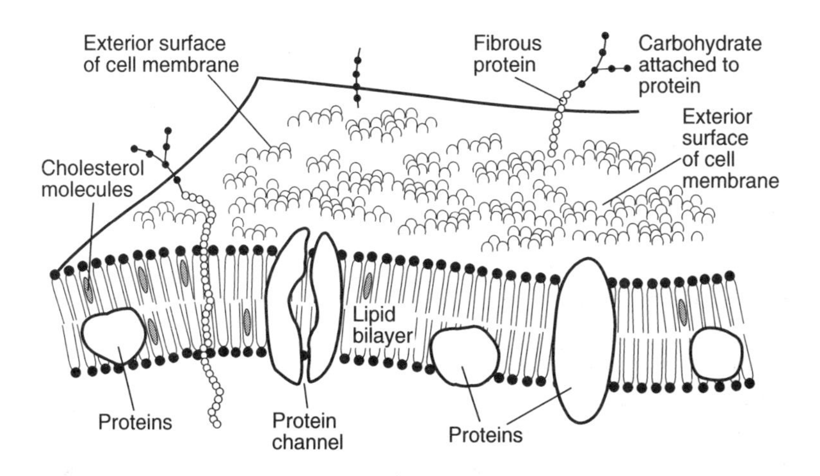

Figure 2.2 The structure of a cell membrane

Each phospholipid molecule is made of a water-soluble 'head' and a water-repellent 'tail'. The molecules line up in two rows: the heads being attached to the watery cytoplasm, point outwards; the tails, being repelled by the water, remain hidden inside the membrane. This arrangement is both stable and flexible. It is stable because the opposition between the phospholipids' water-soluble heads and water-repellent tails prevents the molecules turning around so that their tails point outwards. The flexibility is achieved because the tails, not being rigid, wave around inside the membrane.

Protein molecules are also important in the functioning of the membrane. Some sit on its surface while others penetrate it. They help to carry substances across the membrane by combining with them. By escorting the materials through in this way they facilitate the process of **diffusion** by which substances pass from where they are in high concentration to where they are in low concentration. Some molecules such as carbon dioxide, oxygen, and water, can simply diffuse across without any help.

In some cases the cell needs to accumulate a particular substance against a concentration gradient – i.e. the substance has to move from where it is in low concentration to where it is in high concentration. For this to take place a process of active transport occurs, requiring an input of energy to be used in working a pump.

A further means of intake of materials involves a process called **phagocytosis**. In this process, the membrane flows around the substance being taken in, forming a bubble-like **vesicle** which becomes detached from the rest of the membrane. Once inside the vesicle, digestion begins using enzymes made by the cell. Organelles of bacterial and blue–green algal origin could have entered primitive cells in this way. Indeed, this was the first evolved method of ingestion of food and still occurs in many one-celled animals and in certain white blood cells.

Besides being essential selective barriers to the external environment of cells, membranes are also important within cells. They divide the cell into **compartments**, where different functions can be carried out and have many enzyme-controlled reactions. Internal membranes share the same characteristic two-layered structure of the outer membrane, suggesting that they may have evolved from folds of this outer layer. A double membrane maintains the integrity of the control centre of the eukaryotic cell, called the nucleus. As well as chromosomes, the nucleus contains structures and chemicals which help in the manufacture of protein.

A network of membrane-bound tubes forms the **endoplasmic reticulum** and acts both as an internal transport system and as a surface area for the attachment of centres of protein synthesis, called ribosomes. The proteins that are made on these structures are often transported out of the cell. For this to happen efficiently, there is a packaging department in the cellular factory called the **Golgi body**. Here, the materials for export are surrounded by newly formed membranes before they move to the surface of the cell, where they are passed out by a method that is the opposite of phagocytosis.

Cells united

Since the mid-nineteenth century it has been accepted that all living things, apart from viruses, are made up of one or more cells, but the ways in which these cells are organized varies widely. Sponges are made of cells which work largely independently of each other and which are very loosely connected. In more complex forms, cells group themselves together into **colonies** to form tissues and organs and are completely interdependent. Organisms made of many cells have increased their chances of survival because they have greater scope for adaptation due to the division of labour between various parts.

Two distinct strategies have been followed throughout the evolution of plants and animals. The world of the prokaryotes is the microscopic one. They can multiply at a very fast rate and can therefore rapidly colonize temporary habitats. They have developed such varied biochemistry that they are able to survive in extreme environmental conditions: for example in hot springs, in the deepest caves, and in the abyssal zones of the sea. Generally, though, they are not able to construct true colonies, a stage of development which requires co-operation between cells and loss of individuality. In the eukaryotes, the trend has been towards increasing size and with it increasing complexity. Division of labour between cells has led to the development of larger and larger **multicellular organisms** which can take advantage of all the benefits of their size.

Within populations of single-celled organisms most interactions are accidental. Each cell is affected by the presence of the others but the interaction tends to be competitive rather than co-operative; for example, the cells might compete for food.

There are times, however, when individual cells do interact, most commonly during **sexual reproduction**. In **asexual reproduction**, the genetic material in the form of DNA is replicated by simple cell division to produce identical new cells. This is not the case with sexual reproduction, where genetically dissimilar individuals mix their DNA. The offspring will have some characteristics which are different from those of either of their parents. The resulting variation in genetic makeup is an essential element of evolution because it allows adaptation to selective pressures of the environment to take place. In prokaryotes sexual reproduction is very simple. Two cells of different mating strains form a bridge of cytoplasm through which the DNA of one passes into the other. This is called **conjugation** and also takes place in some eukaryotes. More usually, sexual reproduction requires specialized sex cells called **gametes**.

The first step to the development of multicellular organisms probably was the organization of individual cells into colonies where there was some co-operation between the member cells. Few prokaryotes form colonies, although some species of blue–green algae form chains of cells called **filaments**, in which some cells have specific functions. For example, *Anabaena* has cells that can

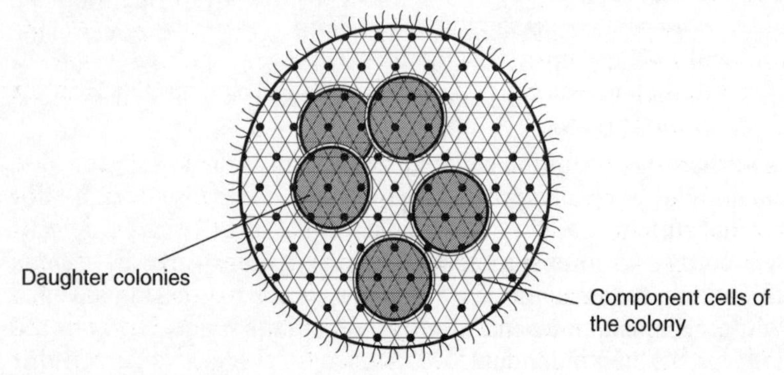

Figure 2.3 Volvox

photosynthesize and also cells which can use atmospheric nitrogen to help make protein. In general, however, it is among the eukaryotes rather than the prokaryotes that co-operation between cells is found.

One of the simplest levels of colonial organization in eukaryotes is that of a group of freshwater algae that includes *Volvox*, a hollow ball of green cells which rotates with the use of whip-like flagella (Figure 2.3).

The cells are separated by a jelly-like sheath although they are connected by strands of **cytoplasm**. Each cell functions separately but the colony as a whole is capable of co-ordinated movement. A few cells are specialized for reproduction. Their activity is synchronized so that at some stages of the cycle they all reproduce asexually, whereas at other stages gametes are produced and sexual reproduction occurs.

A more complex form of colonial organization is found in sponges. These all live either in the sea or in fresh water and consist of collections of cells which form a hollow cylindrical body, open at one end and attached to a solid surface at the other. Simple sponges have up to five different cell types, each performing a special function. Their cavities are lined with flagella which, instead of acting as little paddles as in *Volvox*, draw currents of water through the body, trapping particles of food. Modern-day sponges are the end product of an evolutionary 'experiment', which led to a dead end in which individual cells are both co-operative and competitive. They co-operate to create an overall structure which filters food out of the water but compete with one another for food. Only in true multicellular organisms is there complete co-operation and no competition between cells. The true multicellular organisms did not evolve from sponges.

A popular theory for the origin of multicellular types suggests that a colony of cells, rather like Volvox, grew inwards to form a two-layered structure with a central cavity. It is thought that the cells of this colony became specialized to form **tissues** (groups of cells performing the same functions). Such a specialization may have led to the cells becoming increasingly dependent on each other until they lost all their individuality. Evidence for this idea is the fact that multicellular animals go through this stage during their embryonic

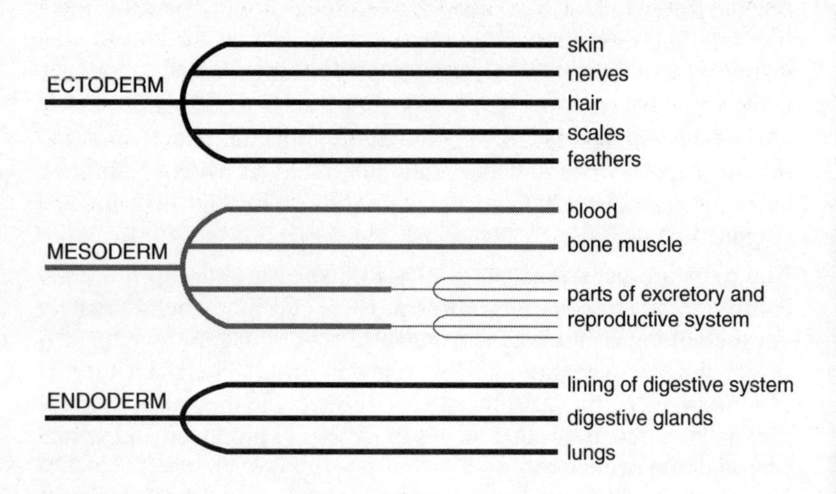

Figure 2.4 The primary germ layers from which all tissues develop in animals

development – **germ layers** (see Figure 2.4). The zygote formed by the fusion of gametes divides to form a hollow ball of cells which becomes invaginated. Further development and cell specialization then gives rise to the adult form.

The simplest true multicellular animals with cells firmly joined together are members of the group called **Cnidaria**, which include the jellyfish, corals and sea anemones. They consist of two layers of cells surrounding a central digestive cavity. Their cells are arranged to form tissues and co-ordination is achieved through a network of **nerve cells**. All cnidarians are aquatic and capture their prey with specialized stinging cells on their tentacles.

Unlike the cells of sponges, the individual cells of cnidarians depend on one another for survival, and their actions are co-ordinated. For example, when prey is captured by one of the tentacles, this information is relayed to the rest of the animal via a nerve net. Simultaneously, cells of the other tentacles contract and draw the prey to the mouth.

More complex than the organization of cells into tissues is the organization of tissues into **organs** and **organ systems**. Organs are distinct parts of the body which perform specific functions and are usually made of several different types of tissues. Large organisms may have the advantage of greater strength and speed but their size also produces problems of obtaining enough food and oxygen to release enough energy. To overcome these they have developed specialized organs like a digestive system, lungs, blood vessels and a heart.

One of the prime needs of larger organisms is a supply of oxygen to their cells. In simple types such as sponges, oxygen can diffuse into the cells from the environment but this is not possible for organisms that consist of many layers of cells. The problem increases with size because the larger and more active an animal is, the more oxygen it needs. Fairly small, inactive animals, such as earthworms, can obtain sufficient oxygen through diffusion over the entire moist skin surface but larger animals have evolved specific organs such as gills or lungs for the uptake of oxygen and removal of waste carbon dioxide. **Gills** and **lungs** have developed to increase the surface area for the exchange of oxygen and carbon dioxide. Blood is circulated between the gills or lungs and the tissues by the pumping action of the **heart**.

In addition to oxygen, animals need fuel from which to release energy. The fuel is processed in the digestive system which breaks down food into simpler chemicals that can be utilized by the body. Waste products of the associated biochemical reactions are dealt with by an **excretory system**. An increasingly complex system of organs requires co-ordination and this is the function of the nervous system. Through rapid response to changes in the external or internal surroundings, an animal's behaviour enables it to survive in its environment.

This principle also holds for plants. One of the most obvious differences between animals and higher plants is that animals move in search of their food at some stage in their lives, whereas higher plants are rooted to one place and produce their own food by photosynthesis. However, despite their static state, such plants do respond to changes in their surroundings such as light, gravity and touch. Shoots grow towards light and away from gravity but roots carry out the opposite responses.

While the organization of cells varies widely between different forms of life, the function of this organization is the same – to meet the particular requirements for each organism's survival. From the unicellular structure of the simplest organisms to the complexity of interdependent organs, specialization through division of labour has enabled the development of the biodiversity we see today.

Guts but no backbone

Imagine that space shuttles, supersonic aircraft, hovercraft, ocean liners, sailing ships, railway trains, and all the other modes of travel that human ingenuity could ever design appeared during a ten-year time span in the fifteenth century. Today's historians would be queuing up to publish their explanations of how such a variety of creativity came about in such a short space of time. That is a fair comparison with the attitude of palaeontologists to the explosion of evolutionary activity that took place in the Cambrian period (543–510 million years ago). Late in 1993, a team of geologists and palaeontologists suggested that in just ten million years, the time it took for an evolutionary sneeze, living forms erupted in an orgy of innovation that far surpassed anything Earth had witnessed before or has experienced since. During the preceding three thousand million years, the 'best' that evolution could come up with were the equivalent of dugout canoes in our transport analogy. There were algae, some flatworms, and the mysterious Ediacara fauna. The latter were named after the Ediacara Hills of the Flinders range in South Australia where their fossils were first found. The Australian geologist **R.C. Sprigg** discovered the first of this weird fossilized zoo in 1947. These creatures dominated the Earth for tens of millions of years, originating about 680 million years ago. It took some extraordinary geological circumstances to make it possible for these organisms to leave any remains at all because they were all soft bodied, without any skeletons whatsoever. The Ediacara Hills provided the right conditions at the right time, their fine sandstone sediments allowing the outlines of soft jelly-like bodies to be preserved. Proving that the Ediacara fauna were not just a 'one-off freak', subsequent discoveries have shown that their distribution is world-wide. According to the dating of some fossils found in Namibia in the late 1990s, they lasted until the Cambrian period. They were related to modern-day invertebrates (animals without backbones) but not resembling any survivor very closely by way of their external features. Indeed, there are some scientists who believe that the Ediacarans were neither animal nor plant but that they showed features of both groups.

Then, without warning, at least little that left a fossil record, evolution burst forth like the proverbial bat out of hell, relative to the normal geological way of doing things. Early in the Cambrian period (about 535 million years ago), a large number of highly organized and well defined invertebrate groups suddenly appeared. They were the first to have heads, middles, rear ends, segments and guts. Some had four legs, some a dozen. Blood, shells and antennae also first appeared. In short, almost all the body forms familiar in modern invertebrates, and several more that have long since disappeared, abruptly materialized in this one great release of evolutionary energy. For this reason, the Cambrian is sometimes called the **Age of Invertebrates**. It was as if a predator had a chance to evolve, initiating an evolutionary arms race in which everyone had to get larger, better armoured or cleverer to stay alive. Perhaps the amount of oxygen in the atmosphere increased, allowing the potential for the development of more active animals. Whatever kick-started the Cambrian frenzy, it was probably easier for new body plans to be successful then than it is now. The world was empty ecologically, with much room for experimentation in niche exploitation. Animals could take advantage of every opportunity. Without competition, there is truth in the Chinese proverb:

> When there are no fish in the pond, even the shrimp is big.

Then it all ended as suddenly as it had started. In the more than 500 million years since the end of the Cambrian, evolution has rested on its laurels, content with variations on old themes. In explaining this phenomenon, palaeontologists face a tough task – after all, they are separated from their subject by 500 million years.

In the mid-1990s three scientists in California came up with an ingenious explanation for the arrival of so many diverse body plans in such short a time. They put forward the hypothesis that there appeared, at this time, cells that are held in reserve during **larval** development and which are later switched on to create the adult body form. These scientists believed that larval development was the key that opened the Cambrian's evolutionary toolbox. For their evidence they looked towards the embryology of marine invertebrates, many of which undergo **indirect development**. Most sea urchins and bivalve molluscs, for example, begin life as a larval stage consisting of a few thousand cells. None of these cells divide more than about twelve times, except for a few 'set-aside' cells that play no part in larval development. The adults, which look nothing like the larvae, develop from these specialized cells once they are activated. It is possible that the first multicellular types resembled the larval forms but, because of their soft bodies, would have left few traces as fossil evidence. They could have reproduced without reaching 'adulthood' (see 'neotony', page 68). The crucial advance, according to the researchers, was the evolution of genetic programmes to switch on and direct animals' complex patterns of development. These are controlled by a large group of **homeotic genes**, which are products of hundreds of millions of years of evolution. Rudimentary versions could have arisen by mutations of a small number of ancient developmental genes.

Whereas Precambrian rocks virtually lack fossils, Cambrian rocks are often crammed with them. The Precambrian era was a period of intense mountain formation, retreating seas, deteriorating climate and an associated Ice Age. The early Cambrian saw a climate improvement and a gradual rise in sea level, leading to progressive flooding of the continental platforms. Previous major upheavals of the Precambrian would have eliminated many species but the return to stability would have suited the survivors and allowed them to evolve in different **ecosystems**. New habitats became available to them by the rise in sea level. Many divergent groups appeared at this time, with hard supporting structures made of silica, calcium carbonate, chitin or phosphates in the forms of shells, plates or spines. The Cambrian seas must have swarmed with drifting jellyfish-like creatures whereas the shallower, warmer waters were exploited by the coral reef builders. Today, corals and their relatives are still abundant with at least 10 000 known species. All of them are aquatic.

As a contrast, in the post-Cambrian world, competition is so severe that if you have not perfected the art of making a living, selective pressures will snuff you out like a candle. Animal life on Earth has matured; the halcyon experimental days of its carefree youth are long gone. Extinct groups abound as fossils and, therefore, as evidence of evolutionary complacency.

The **graptolites** were one such group. They drifted in the oceans or anchored themselves to the sea bed throughout the Ordovician and Silurian periods (505–410 million years ago) but are now totally extinct. Their delicate branching colonies have left fossils that resemble small saw-like structures embedded in rock. Once considered to be colonial relatives of the corals, like modern-day sea firs (Hydrozoa), they are now thought to be closer to the ancestors of vertebrates (animals with backbones).

Worms

A number of invertebrate groups are collectively called worms but may be only distantly related to each other. Worm-like animals were probably among the earliest invertebrates and are known from as far back as the earliest geological period (Precambrian) but, being soft bodied or microscopic (or both), they have not survived well in the fossil record. Tooth-like fossils made of phosphate, called condodonts, are thought to be fossilized mouthparts of worms by some palaeontologists.

True segmented worms like earthworms are in a group called the Annelida (*annulus* = 'ring') and are sometimes found as fossils but, more commonly, the tubes they made and inhabited on the sea bed are preserved. Today there are well over 7000 known species including mostly marine worms, leeches and earthworms.

In rocks of the Cambrian era, fossils have been found which seem to link the Annelida with a group of jointed legged invertebrates, the **Arthropoda**. Such a fossil is ***Aysheaia***. This small marine animal was segmented like the annelids but had walking legs, more typical of arthropods like the centipedes of today. Their descendants invaded land and possibly gave rise to the present-day relict species ('living fossil') called ***Peripatus***. It closely resembles its Cambrian ancestor and survives today as 70 species throughout parts of tropical America, Africa and Australasia. As long ago as 1874, it was decided that *Peripatus* was more closely related to arthropods than to annelids when **H.N. Mosely** became the first qualified naturalist to examine living specimens. Ironically, they were found

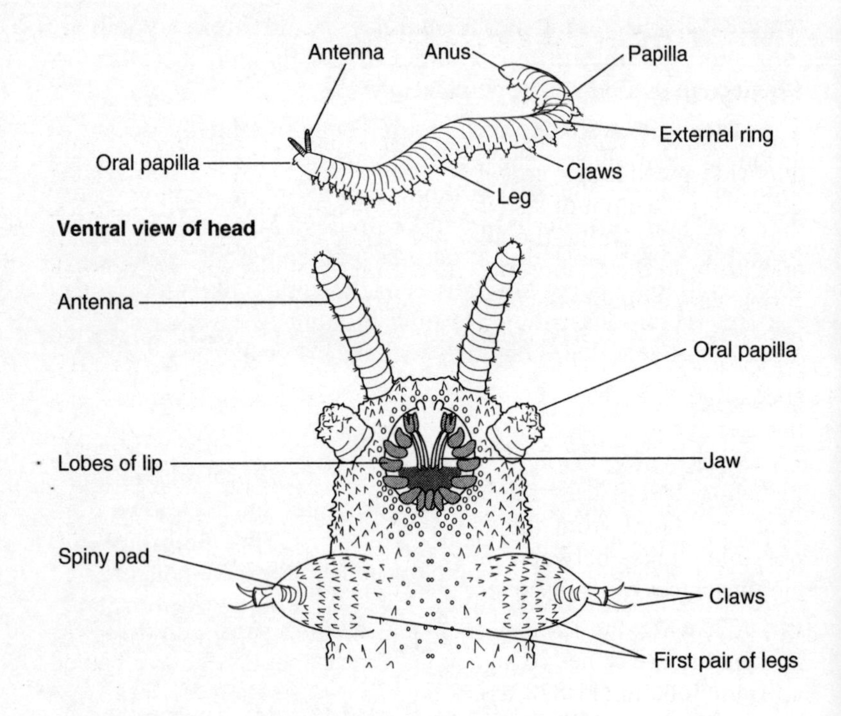

Figure 2.5 Peripatus, the 'velvet worm'

in a forest habitat in South Africa even though Mosely was on an oceanographic survey at the time, on HMS *Challenger*! He took advantage of the ship's seven-week refit at Cape Town and explored the local area. Their fossil ancestor, *Aysheaia*, is over 500 million years old. Although nineteen specimens of this fossil have been found, the first was little more than a rusty-coloured stain in a piece of limestone. Yet it was enough to convince the specialists that the structure of its body and legs relate it, albeit distantly, to *Peripatus*, commonly known as the 'velvet worm', which is still crawling around in some damp tropical forests at this moment (Figure 2.5).

The armoured divisions

Shells and spines

The **molluscs** (snails, clams, squids and slugs) are highly successful and widespread today, probably because their ancestors evolved the effective protection of a shell. One of the most primitive molluscs, which may resemble the ancestor of the whole group, is ***Pilina***, appearing in the Cambrian with a simple cup-shaped shell. Such forms were thought to have become extinct about 350 million years ago but in 1952 living specimens of a species (to be called ***Neopilina***) were dredged from the depths of the Pacific. To the non-specialist, *Neopilina* is a somewhat boring looking animal – just another limpet about 2 cm in diameter – but, in the words of Professor **C.M. Yonge**, one of the most famous British malacologists, it was 'the most exciting molluscan event of this century'. Apart from the fact that it is 'living fossil' showing possible links between annelids and molluscs, the whole circumstances of its discovery are astonishing.

In 1952, when the Danish deep-sea research ship *Galathea* was nearing the end of her voyage in the Pacific, her dredge was hauled up from 3600 m (11 878 ft) off the coast of Costa Rica. In it were ten living limpet-like animals and three empty shells. When eventually these were examined in detail at Copenhagen, they were found to belong to a species of molluscs which reputedly had died out 350 million years ago. Nothing quite like it had been seen alive before – and, because it had been brought up from the deep ocean

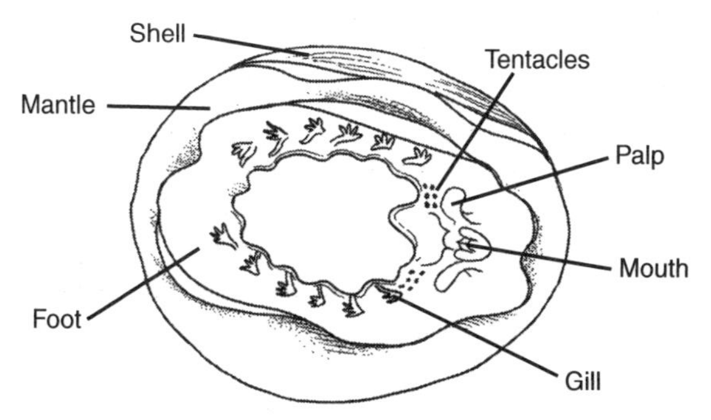

Figure 2.6 *Neopilina*, ventral view

bed, it was not possible to do more than speculate about its way of life. Six years after this particular species of *Neopilina* had been found off the coast of Costa Rica, another species was caught off the coast of Peru. Yet another species was found off the coast of California in 1962 and then, only nine years later, a fourth species was collected in the Gulf of Aden at a depth of 2700–3500 m (9000–11 850 ft). From these data it is fair to assume that these animals have a very wide distribution in the deep oceans of the world. The amazing point about the finding of *Neopilina* is that it had remained undiscovered for so long. Since 1872, with the three-year epic voyage of HMS *Challenger* that marked the birth of oceanography, there have been scores of voyages by deep-water research vessels, some covering vast areas with intensive dredging. Eighty years of searching had missed it completely.

By the Ordovician period (505–433 million years ago) all the present-day known groups of molluscs had evolved in the oceans. The fossils of their ancestors show striking variations on the original primitive body plan shown by *Pilina*. Fossil records of molluscs owe their abundance and completeness to their well defined solid shells, ranging from a few millimetres to the Cretaceous type ***Inoceramus***, which was well over a metre across, like today's giant clams.

In the snail-like class, **Gastropoda**, the shell is in one piece and is often spirally coiled. Fossils are as old as the Cambrian and it is likely that they began to invade land in the Cretaceous period (144–65 million years ago).

Whereas most molluscs are relatively slow moving and some are almost stationary as adults (e.g. clams), one group developed a form of jet propulsion by which they expel water with enough force to push themselves through the water very quickly. These types are the **cephalopods** and include the present-day squid, octopus and nautilus. Their ancestors were the **ammonites**, which were abundant in the oceans of the Palaeozoic and Mesozoic periods. They are named after the Greek god Zeus Ammon, because of their resemblance to the god's spiral horns. Thousands of extinct species are known by their fossils, which may be up to a metre in diameter, like ***Titanites***. Ammonites are well represented in the Jurassic (202–144 million years ago) and Cretaceous rocks, as are the

belamnites. These are bullet-shaped remains of chambered shells which were the internal skeletons of squid-like creatures, much the same as the cuttlefish has an internal shell.

Towards the end of the Cretaceous period the ammonites, belamnites and most nautiloids became extinct. This was probably at the same time as the extinctions of many other groups. Interestingly, there is no universally accepted explanation for this world-wide mass extinction which, in geological terms of time spans, was very sudden and catastrophic (see page 184).

Not all shelled forms were molluscs: filter-feeding **brachiopods** (lamp shells) were already present in the Cambrian and remain today but in relatively very small numbers and species diversity. It was among the Palaeozoic and Mesozoic fauna that they reached their evolutionary peak. Brachiopods resemble bivalve molluscs superficially but their internal anatomy is very different. Fossil brachiopods have been used extensively for determining geological zones and conditions in the remains of prehistoric seas. Today about 300 species of living lamp shells are known (compared to 30 000 fossil species). They reached a peak in the Ordovician period when they must have been as numerous as molluscs are today.

The external plate-like skeletons of the spiny-skinned invertebrates, the **echinoderms**, allowed this group to be particularly well preserved as fossils. Living forms include the familiar starfish, sea urchins, brittle stars, feather stars and sea cucumbers. About 20 000 species are known, of which about 5000 are still in existence. They seemed to reach a peak of abundance in the Ordovician. Fragments of such fossils are common in many limestone deposits. By the Mesozoic, sea urchins became abundant and whole developmental series of these have been found in successive layers of rock, tracing their evolution.

The first joints

The most successful group of invertebrates is undoubtedly the group called the **Arthropoda** ('jointed legged'). They have been described as the group 'beyond census', simply because there are more of them than any other kind of animal on this planet. They include familiar creatures such as insects, spiders, scorpions, millipedes, centipedes, mites, crabs, lobsters, prawns, and a vast number of types which have no common name. To the collection may be added a number of totally extinct groups, of which the best known are perhaps the **trilobites**.

There are more species of spiders and their relatives (**arachnids**) than any single-celled animals. Crabs and their relatives (**Crustacea**) are both more numerous and diverse than fish. **Insects** outnumber all other arthropods. About a million species have been catalogued and many more remain to be discovered and described. In fact, about three-quarters of all known animals are insects.

The earliest arthropods are considered to have evolved from segmented worm-like animals but not surprisingly there is little fossil evidence of such delicate, soft-bodied animals. Nothing is really known of the Precambrian history of the arthropods. However, Cambrian deposits contain many different animal groups with hard skeletons. It appears that the development of hard **exoskeletons** (skeletons on the outside of the body) was widespread in this period, but their origin continues to be a controversy among palaeontologists. Trilobites, the oldest arthropods (Figure 2.7), appeared about 570 million years ago and were all extinct 300 million years later. They thus span nearly the whole of the Palaeozoic era. Fossil trilobites are always found with fossil marine animals so we know that they must have lived in the sea. Most lived on the sea bed and a number of them have been found rolled up like modern-day wood lice or slaters. Some were tiny – no more than 0.5 mm in length – but the largest are nearly 1 m long.

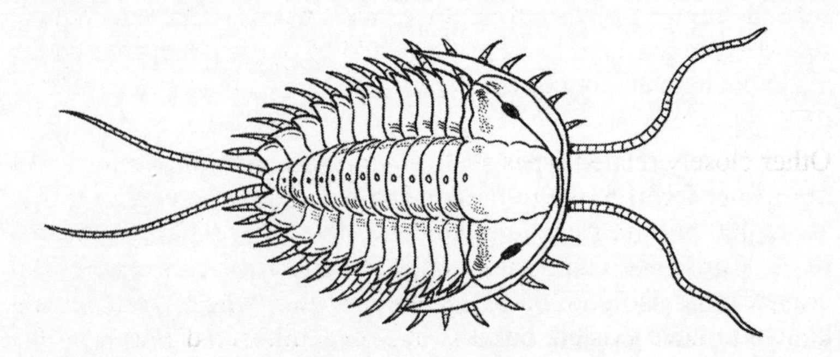

Figure 2.7 A trilobite

Another spectacular group of extinct arthropods were the **eurypterids** or 'sea scorpions' (Figure 2.8). They are thought to be the progenitors of the spider group of arthropods (Arachnida). Some reached an enormous size, at least 2 m in length. They had powerful jaws and were probably fierce predators. They first appeared in the Ordovician period, about 100 million years after the first trilobites, and similarly died out in the Precambrian.

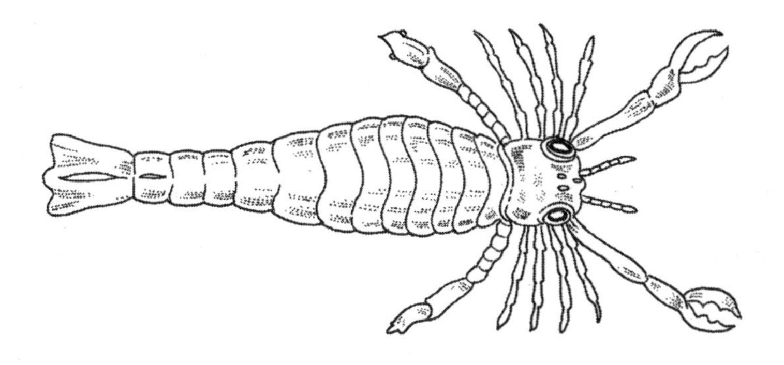

Figure 2.8 A eurypterid

The marine arthropods were among several groups of ocean dwellers that suffered extensive extinctions at the end of the Palaeozoic era (about 245 million years ago). One of the possible reasons is that at this time the continents had drifted together to form a single 'super-continent', called Pangaea. The consequent reduction in the area of sea shore would have greatly decreased the available habitats for coastal marine animals and many would have died out.

Other closely related types such as scorpions and mites managed a more successful move to the land in Silurian to Devonian times, about 400 million years ago. Spiders and millipedes first appeared in the Carboniferous period about 300 million years ago. Fossil crustaceans also date back to the Cambrian when barnacles are known to have existed, but it is only in sediments dating from the Triassic period, some 300 million years later, that we find fossil crabs and prawns.

Have backbone – will evolve

Fishes were the first animals with backbones. None are able to control their body temperature but all are able to breathe oxygen dissolved in water using gills. Their fossils first appear in marine 450-million-year-old Ordovician sediments. Since that time they have diversified to inhabit almost every part of the planet where water exists. There are an estimated 23 000 species of fishes surviving today. Three out of every five vertebrate species are fishes. Their origin is difficult to ascertain but basically their progenitors developed along two pathways – one developed biting jaws and the other developed a type of rasping sucker-like mouth. Where both jawless fishes (**Agnatha**) and jawed fishes (**Gnathostomata**) occur as fossils, they each already have anatomical features identifiable in modern-day fish. We have seen that invertebrates rarely fossilize well unless they have some sort of hard bits and it is unlikely that the direct ancestors of fish had any.

In seeking the link between the spineless and the spined, evolutionary biologists have found an animal which may appear most unlikely to many – the humble 'sea squirt'! The larval form of this group of invertebrates is thought to have been the blueprint for all animals with backbones, including us. Like a real-life Peter Pan, the sea squirt never really grew up but developed the ability to reproduce while in the larval stage. This phenomenon is called **neotony** or **paedogenesis**. The larval form looks like a mini tadpole but has features which are present in primitive fish – for example, gills, a ventral heart, and a dorsal hollow nerve cord. It also has a rod (notochord) of skeletal tissue running along the dorsal surface. This notochord is the precursor of the vertebral column or backbone and is present in every vertebrate at some stage in its life history. Its presence makes the tadpole larva an effective swimmer. Several types of present-day amphibians show a similar process of neotony because they too can reproduce while in their larval stage. A good example is the axolotl, which changes into its adult form, the tiger salamander, only in certain specific environmental conditions but can reproduce when in its equivalent 'tadpole' stage.

In 1774 a strange little animal was picked up on the coast of Cornwall, England. It was subsequently preserved in alcohol, and

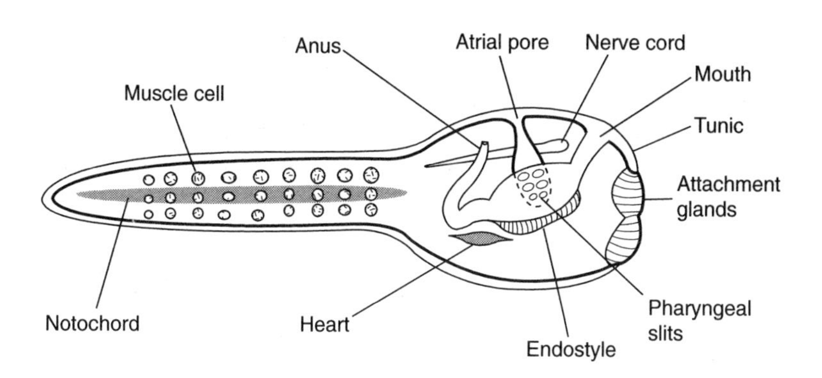

Figure 2.9 The tadpole larva of a sea squirt

sent to the celebrated German naturalist **Peter Simon Pallas**. There seems to be no record of who picked it up or why it was sent right across Europe when there were so many competent naturalists in Britain who might have examined it. Subsequently, Pallas described it in a footnote of a book he was publishing, giving it a very brief description and naming it *Limax lanceolatus* under the impression that it was a sea slug. Half a century later, on 21 December 1831, **Jonathan Couch**, one of the leading English naturalists of that time, was walking along the shore near Polperro, in Cornwall, after a storm. In the usual practice of curious naturalists, Couch was beachcombing and, after turning over a pebble about 50 ft from the ebbing tide, he saw a tiny tail sticking out of the sand. He dug out the complete animal and took it to his marine aquarium where he could examine its habits. Couch sent the specimen to **John Yarrell**, a fish specialist, who in 1836 described it in his book *A History of British Fishes*. He called it *Amphioxus*, which means 'sharp at both ends' and considered it to be a primitive type of fish. The common English name for it is the lancelet. Although now rare around the beaches of Britain, on the coast of Amoy, China, it is common enough to be used as food. In the 1960s over 35 tonnes a year were harvested but over-fishing has reduced the catch considerably.

The relationship of the lancelet with the rest of the animal kingdom remains one of the most interesting features of this rather drab little animal. It resembles a 'stripped-down' vertebrate in having a dorsal nerve cord lying above a supporting rod, the notochord, and an arrangement of muscles along its tail like a fish. At the same time it lacks a backbone, jaws – or indeed any bone. It also lacks anything that can be called a brain or eyes and other sense organs associated with a brain. So it cannot be classed as a true vertebrate, yet it is as close to one as any invertebrate could be.

There is a fossil dating from the Cambrian which has been found in the Canadian Rockies and resembles the lancelet in many ways. The site of its discovery is a particular rock formation known as the Burgess Shale, which has provided a harvest of unique fossils, few of which resemble any living forms. The particular area lies in the Yoho National Park on the slopes of Mount Stephen in British Columbia. It has been well known by palaeontologists since 1909 when it was first made famous by **Charles Doolittle Walcott**, who found vast numbers of animals that had been covered by sediments so quickly that their soft bits were preserved for posterity as fossils.

In his book *Wonderful Life,* **Stephen Jay Gould** focuses on the evolutionary significance of this collection of very strange fossils. The one resembling the lancelet is called *Pikaia* and could be the earliest known forerunner of the vertebrates. A prominent strand in the dorsal region of the animal appears to represent the notochord. The current view is that both the lancelet and the vertebrates evolved from the same ancestors as did the sea squirts. Both lancelets and sea squirts feed in the same way but most sea squirts are anchored to a solid support when adult and so look most unlike the lancelet.

Jaws – to have or have not?

The first recorded fossils with an unmistakable fish-like appearance are the agnatheans (*a* = 'without'; *gnath* = 'jaw'). Today the only remains of this primitive group are the lampreys and parasitic hagfish. They are considered to be the descendants of a group of armoured jawless fish called the **ostracoderms**, although the main claim to fame of lampreys is that they reputedly caused the death of Henry I of England who died as a result of eating too many of them in one sitting!

(a)

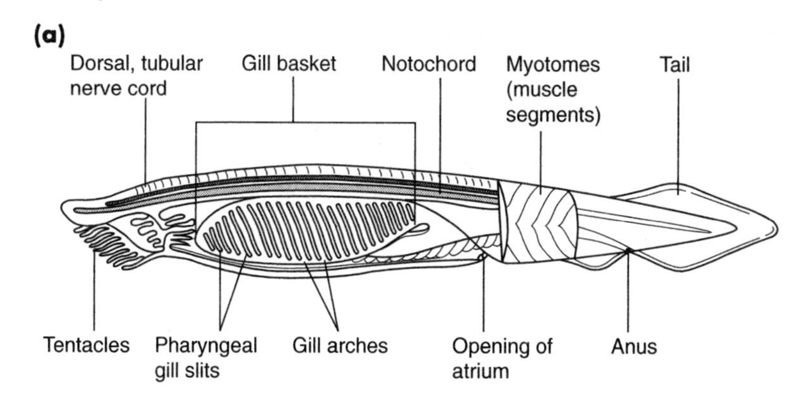

(b)

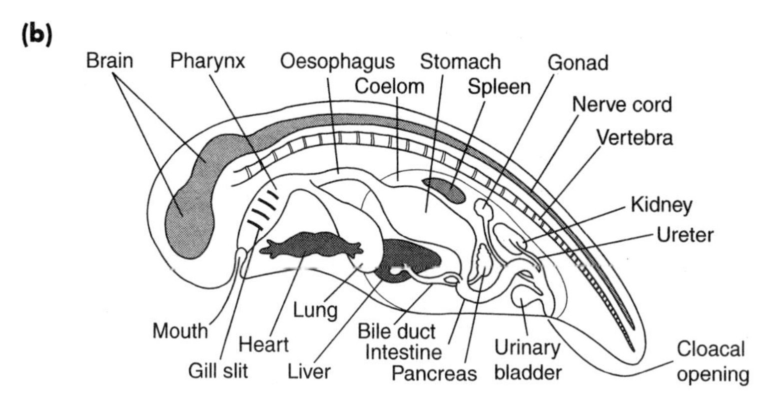

Figure 2.10 (a) Amphioxus branchiostoma; (b) a typical vertebrate

The ostracoderms of the late Devonian were small sea-bed dwellers covered with protective bony plates. They probably made a living by feeding on invertebrates in the ancient sediments. At that time, large predatory crustaceans were competing with them for food. Apart from not having evolved the biting jaws of modern-day fish, they also lacked paired pelvic fins. By the Silurian and Devonian they had diversified but had largely died out by the mid-Carboniferous. They were probably at a disadvantage when it came to competing with the newly evolved jawed fish – the gnathostomes.

The two lineages for the evolution of the jawed and jawless fish are completely separate, as studies of details of the gill structure have confirmed. The lack of early gnathostome fossils in Ordovician deposits may be due to their evolution in a deep-sea environment. It was not until the Silurian age that they first appeared and by the end of this period a vast array of different forms had developed. They had two sets of paired fins, pectoral and pelvic.

Both jawed and jawless fish were more active than the other animals which were alive at the same time. Their internal skeleton gave them scope for greater development of muscle attachment for versatility of movement. Activity requires energy and the more active an animal is the more selective pressure is exerted on its metabolism. An animal's metabolic rate is an expression of the efficiency of three internal factors:

- The surface for gas exchange
- The feeding mechanism
- The excretory mechanism

The greater the metabolic rate, the greater the amount of waste there will be to be eliminated. Gills for gaseous exchange, jaws for obtaining food, and kidneys for excretion are all well developed in fishes and were the prerequisites for the evolution of the variety that are seen today.

A simple division, based on the material making up their skeletons, separates the two groups of true fish-like animals which exist. One major group (the **cartilaginous** fish) has **cartilage** (gristle) making up the skeleton and includes the sharks, skates, and rays. The rest have skeletons made of bone.

Soft skeletons

There are over 800 species of cartilaginous fish living today but few fossils of their ancestors survive because of the soft nature of cartilage. Although most are marine, a few have penetrated fresh water and this may be due to their peculiar method of coping with the problems of living in salt water. Water tends to be lost through fishes' cell membranes to the sea because of the sea water's higher salt content. It leads to dehydration by **osmosis** or diffusion of water out of the fish. Water diffuses from where it is in relatively high concentration to where it is in relatively low concentration until dynamic equilibrium exists. At this point there is an equal rate of movement of water in and out of the fish. That is why slugs shrivel up when you put salt on them in your garden!

The prevention of dehydration has been of major importance in the evolution of many groups of animals and plants. How do marine fish cope with the tendency to lose water by osmosis? One of two possible strategies are used. Bony fish have special salt-secreting glands in their gills so they can 'drink' sea water and maintain a concentration of body fluids which is the same as the surroundings, thereby allowing water to pass across their cell membranes at the same rate in both directions. Cartilaginous fish do not use this method. They retain their waste urea in their body fluids to achieve the same end result – i.e. the concentration of their fluids becomes the same as the surrounding sea water. This adaptation to marine life may have prevented cartilaginous fish from exploiting fresh water to any great extent. They also lack swim bladders which are found in most bony fish and so lack natural buoyancy, making it necessary to swim continuously or sink.

Bony fishes

Today's dominant fish are those with bony skeletons. They thrived and diversified in the Devonian, first in fresh water and then in the sea. The end of the Silurian saw a sudden and massive blossoming of fishes. This expansion coincided with their colonization of fresh water, which gave them the opposite problems to those of living in sea water: more water would tend to enter the fishes than would leave by osmosis. Kidneys are used to eliminate large amounts of very dilute urine and lost salts are regained by active uptake across the gills.

Fresh water provided more environmental selective pressures than sea water because of its greater potential for variation in physical properties. It therefore provided greater opportunity for the development of species diversity. Temperature can change quickly, levels of nutrients can vary widely, and oxygen concentration can diminish rapidly. The marine environment tends to be more stable and less stressful in the context of these factors.

Various primitive bony fish living today suggest one response to the stress of decreasing oxygen supply. They include the reed fish *Polypterus*, the gar pike *Lepisosteus* and the bowfin, *Amia*. They can all breathe air with special throat pouches – a useful adaptation for life in stagnant water in which oxygen levels can fall to almost zero.

The ability to breathe air was taken a few stages further by the **lungfishes**, which now exist in three different continents as different species. Those living in Africa and South America can withstand conditions of total drought and, together with their Australian relatives, are confined to regions that have seasonal (or longer) droughts. As the land begins to dry up, they burrow into the mud and spin a water-retaining mucous cocoon around themselves. They continue to breathe air until the rains arrive. While in the cocoons their excretory metabolism changes so that they produce urea rather than their normal poisonous excretory product, ammonia. Only with the arrival of the rains will there be enough water to dilute the ammonia and render it harmless. At one time it was thought that lungfish were the ancestors of amphibians but it is now thought that amphibians were more likely to have evolved from ancestors of a different group.

Ancestors of most of today's species of fish were ray-finned, with fan-shaped fins. Besides their familiar characteristically paired fins, many have swim bladders for regulating depth and modified jaws for hunting a variety of prey. The most advanced fish have pectoral fins on their cheeks and pelvic fins almost under their chins. This arrangement gives even greater locomotory ability.

The remains of another day

In 1938 an ancient lobed-finned fish, thought to have been extinct for 65 million years, came back from the past when a living specimen was caught in the Indian Ocean. This catch of the century was the, now famous, **coelacanth** and its discovery was analogous to finding a living dinosaur. Its name literally means 'hollow spines', referring to those of the fins. The fossilized remains of coelacanths are well known in sedimentary rocks from 400 million to 65 million years old, but they have not been found in rocks younger than this. Their fossil record mysteriously ends at about the time of the demise of the dinosaurs. So it was generally believed that coelacanths were extinct, having been carried away in the long Cretaceous cortège. In 1938, however, ill fate befell an unlucky specimen of a 1.5 m (5 ft) long, 57 kg (127 lb) ugly, bluish-grey fish with strange lobed fins. It had strayed a little too near the trawler *Nerine* and was netted off the Comoros Islands at 32 m (120 ft), among a 3 tonne catch of other fish. Fortunately, for science but not for the fish, it was not thrown back to the sea but was kept by a keen-eyed fisherman with an interest in the unusual as well as in food. Ironically, this evolutionary prize almost never made it into the history books due to the timing of its capture. It was three days before Christmas when it was landed at East London, South Africa, and was almost overlooked by the curator of the local museum, **Marjorie Courtenay-Latimer**. She was in the habit of visiting the dockside frequently to collect unusual creatures from fishermen, but when Captain Goosen, the skipper of the *Nerine*, rang her office at 10.30 a.m. on 22 December, she is quoted as saying 'I at first thought, what shall I do with a fish now? So near Christmas? Then I considered I should go down and wish the men on the trawler a Happy Christmas.'

The new fish was not recognized as a coelacanth at first but clearly it was a specimen to be preserved. She thought it was 'a lung fish gone barmy!' However, its size and smelly condition, combined with the inadequate facilities at the museum for dealing with large specimens, posed a problem. Superimposed on this was the deepening lethargy of a hot Christmas holiday period. Indeed, getting the fish from the docks to the museum was almost like a comic opera, with a reluctant taxi driver eventually being persuaded to take Miss Latimer and her festering fishy friend back to the museum. Thus, unrecognized for what it was, wrapped in old sacks, the first extant specimen of a lineage supposedly extinct for 65 million years entered the annals of contemporary ichthyology. The nearest practising ichthyologist at the time was Professor **James L.B. Smith**, at Rhodes University in Grahamstown, 400 miles away. Miss Latimer sent him a letter with an accompanying sketch but it did not reach him until January. Preservation of such a large specimen presented what was to be an unsolved problem. The local mortuary refused to accept the corpse in its cold store so finally, a local taxidermist, Mr R. Centre was given the job of preparing the fish for posterity. It is said that he had very little experience at stuffing fish of that size and it ended up in formalin in a makeshift piscine sarcophagus. On 26 December an examination of the fish revealed alarming signs of deterioration because the formalin had not penetrated the internal organs. A 1.5 m coelacanth decomposes with as much enthusiasm as any other similar-sized fish. The practical consequences of decay dictated the next step and so most of the internal organs were discarded and dumped. On 3 January, Marjorie Courtney-Latimer received a telegram from J.L.B. Smith. It read 'MOST IMPORTANT PRESERVE SKELETON AND GILLS = FISH DESCRIBED'. A frantic search through smelly rubbish bins ensued but by then the rotting parts had gone where all decay leads, especially in those temperatures. In addition to this disaster, it was then found that the original photographs of the freshly caught fish had been spoiled – and, to add to the fiasco, the museum authorities had mounted the skin before any further examination could be carried out. On 16 February Professor Smith was eventually able to inspect the now eviscerated and badly preserved carcass. He recognized it as a member of the lobed-finned fishes, the Crossopterygii and dubbed it 'Old Fourlegs' because of its lobed pectoral and pelvic fins, which he thought enabled the fish to creep along the sea bed. Smith published a book on the subject with *Old Fourlegs* as its title.

The fish was named after its discoverer and is called *Latimeria chalumnae* (*Latimeria* after Miss Latimer; *chalumnae* after the mouth of the river Chalumna, where the fish was caught). At the age of 41 Professor Smith had made history. It had taken a little longer for *Latimeria* – 65 million years! For nearly 14 years Professor Smith encouraged and organized searches for more specimens but for various reasons never continued his anatomical studies in great detail. These were carried out, in meticulous detail by Professors Millot and Anthony of the Paris Natural History Museum, and were based on several of the many specimens collected after 1952 from the archipelago of the Comoro Islands, a French colony north-west of Madagascar (now Malagasy). Smith had flooded East Africa with leaflets in English, French, and Portuguese, offering a reward of £100 for the first two specimens caught. On 20 December 1952 one particular coelacanth hit the headlines with considerable media coverage when it almost caused an international incident. In what must have been the most remarkable reason for the invasion of another country's sovereign territory, the South African president authorized a military airborne operation to kidnap a fish – a coelacanth caught in waters which legally were under the possession of the French government! They were very gentle about it and there was no bloodshed!

Since the 1960s many specimens have been caught and they have become valuable trophies for scientists, museums and the Comoran government alike. Indeed, Francois Mitterrand, one-time French president, was presented with one when he was in office. Some 200 of these 'official' coelacanths are on record. By the 1980s, as the annual catch rose from the traditional three or four to a dozen or more, scientists began to worry that the 'scientific' trade might put the species in danger. Since 1991, coelacanths have been completely protected under the Convention on International Trade in Endangered Species (CITES) In fact, *Latimeria* is still making headlines. In the late 1980s Hans Fricke, exploring from a submersible craft, filmed living coelacanths in their natural habitat at a depth of 200 m, and millions of television viewers saw a dying specimen filmed when Sir David Attenborough made his series *Life on Earth* in 1991.

On 25 March 1999 *The Times* carried an article, entitled 'Second "living fossil" found on fish cart'. Mark Erdmann, a marine biologist from the University of California, Berkeley, was on his honeymoon in Indonesia when his wife noticed an odd-looking fish being taken by cart to the fish market in the town of Manado. It turned out to be a cousin of 'Old Fourlegs' himself and has been named *Latimeria menadoensis*. This second species was caught in the Celebes Sea near the Indonesian island of Menado Tua and suggests that coelacanths are more widespread than had been supposed. Local fishermen occasionally catch the fish and sell them for food. It appears that there is a large and healthy population of this second species living in Indonesian waters.

So what is it about this particular 'living fossil' that attracts so much attention? The answer probably lies in the media's intensely exaggerated publicity which made the nonsensical claim that *Latimeria* is a missing link between humans and fish – the ancestor of the first creature to leave water and breathe air! The term 'missing link' is totally inappropriate, but nevertheless the coelacanth is an exceedingly interesting animal. What is meant by 'living fossil'? The term is difficult to define but it could be described as a living organism that does not differ significantly from fossils which have already been found and described. To this category belong the magnolia, maidenhair tree (*Ginkgo biloba*), lungfish, certain molluscs like *Nautilus* and, of course, the coelacanth. In a sense, coelacanths are special because they belong to a group that was known to scientists from fossil remains long before they were discovered swimming in our modern oceans.

With such intense interest in *Latimeria* and in the one small archipelago where it was first caught, it is no wonder that the French authorities placed an embargo on coelacanth specimens, allowing only French scientists to study them. They even went to the extent of placing injunctions on specimens that they donated to various museums around the world, allowing them to be used exclusively for display purposes. Research on the anatomy and physiology of the coelacanth was therefore left in unhealthy isolation to a small number of scientists. It has been firmly established that coelacanths, although fish, possess many features that are shared by land-living amphibians, reptiles and even mammals. However, they are *not* 'missing links' between ourselves and our fishy ancestors, but *are* one of a variety of types of fish that are very closely related to the first fish with limbs (Figure 2.11). The group of fish thought most likely to be ancestral to amphibians are the **Rhipidistia**, which are unfortunately all extinct. But that's what someone said about 'Old Fourlegs' before 1938!

Ironically, the fishermen of the Malagasy region have probably known the coelacanths for hundreds, if not thousands, of years. They even have a local name for it, *Kombessa*, but regard it as a poor food fish unless salted and dried. They say that its skin is slimy, and when caught it oozes oil. By all accounts, long before 1938 they made good use of it by making 'sandpaper' out of its skin to roughen bicycle tyres before mending punctures!

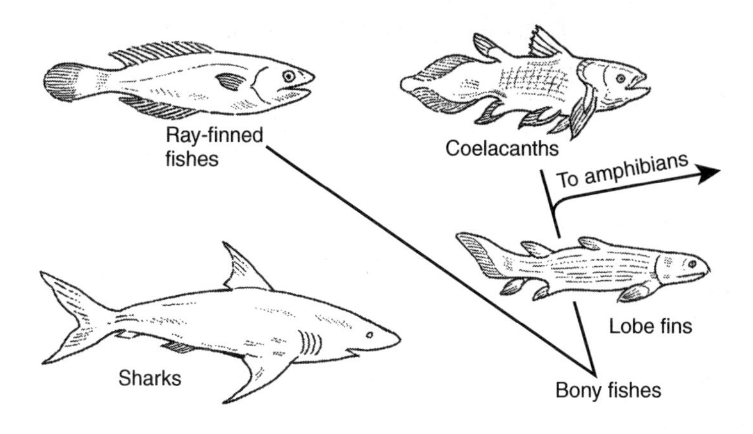

Figure 2.11 The relationship between the coelacanth and other fish

Unfortunately for *Latimeria*, there are now great economic incentives for the locals of the Comoro Islands to catch it – dead or alive. Those who wish to profit by risk and initiative will always find a niche to exploit in our world of materialists. Japanese dealers have been paying the fishermen of the Comoros to hunt for coelacanths because they have the idea that any fish that has cheated extinction for so long must hold clues to everlasting life! Spinal fluid from the fish reputedly sells for hundreds of thousands of yen for a teaspoonful. Black-market bounty will continue to be paid by museums, aquaria, and scientists for new specimens and this will persist until the last coelacanth in the world is eventually caught. For a few dollars more we will have cleared away all the remains of another day.

3 INVASION

The first wave

The edges of lakes, rivers and shallow seas in the humid tropics of the Silurian period (433–410 million years ago) provided an environment that was not very different from water itself. Aquatic plants obtained nutrients from the water and also used it for support and as a medium in which to reproduce.

The fossilized reproductive parts of an ancient **liverwort** provided the first convincing evidence that these cousins of the mosses were the earliest land plants. In the early 1990s British palaeobotanists found, for the first time, a firm link between the liverworts and a particular type of plant spore whose fossil record extends right back to the Ordovician period, some 460 million years ago.

The idea that primitive liverworts were the first land plants was proposed in 1982 but the link was thought to be too tenuous to prove the identity of the plants that had produced the Ordovician spores. Ironically **Dianne Edwards** of the University of Wales, Cardiff, who led the team responsible for the new discovery, was among the critics of the original proposal. Edwards and her colleagues found a 1.5 mm fragment of fossilized plant tissue bearing spores matching those from the Ordovician. The fossil dates from the Lower Devonian period, around 400 million years ago, and consists of spores contained within a forked structure which looks like part of the reproductive apparatus of a living liverwort.

It would have required very little adaptation for small plants to continue to function in the moist atmosphere on the watery fringes. Then, possibly because of increasing competition for space and light and also because of drier climatic changes, some plants were at more of an advantage than others if they could survive in less damp conditions. Like all organisms, they still needed a supply of water but they now had to have a method of absorbing it from the soil. This is a function of **roots**, whereas previously primitive plants could absorb water and minerals over their total surface area. They also needed a way of preventing dehydration and a means of supporting themselves on land. New materials were needed to firmly cement cells together and to keep plants upright so that they could compete for light. Biochemical evolution gave them the ability to synthesize **lignin** and **cutin**, which partly solved these problems. As recently as 1998, evidence has been obtained by scientists at the University of Iowa that land plants may owe their evolutionary success to the accidental duplication of a gene early on in their history. The gene codes for a protein called **actin**, which is a key component of the internal skeleton of cells. Like animals, higher land plants have many different actin genes, and the different versions of the protein have different properties. This diversity helps cells perform specialized functions, an essential step in the evolution of complex multicellular organisms. The taller the plants grew, the further away from the roots were the stems' growing points. Water and minerals needed to be transported to the aerial parts and this necessitated the development of tube-like rigid cells – called **xylem** – in roots, stems and leaves.

The earliest fossil land plant to be accurately dated is ***Cooksonia*** from the Upper Silurian (Figure 3.1). It was more highly developed than any other plant of its day, with recognizable ancestral roots (**rhizoids**) and a waterproof outermost layer of cells (**epidermis**) to protect it from drying up. A **vascular system** of xylem for water and mineral transport, and **phloem** for the transport of manufactured foods, was present. No leaves existed, however, and reproduction was by means of **spores** produced in **sporangia**. There were no flower-like structures to produce seeds.

A close relative of *Cooksonia*, called ***Psilophyton***, still survives among the ancient larval beds of Hawaii (Figure 3.1). Both *Cooksonia* and *Psilotum* are classed as **pteridophytes** (*ptero* = 'winged'; *phyton* = 'plant'), which include ferns, clubmosses and horsetails. It was the ancient classical botanists who coined the term pteridophyte because fern fronds reminded them of angels' wings.

Their stiffened stems and primitive roots encouraged erect and firm growth. Throughout the Devonian period (410–360 million years ago) the landscape altered spectacularly. Bare rocks were quickly covered with miniature forests which were able to support small plant-eating animals like springtails and millipedes.

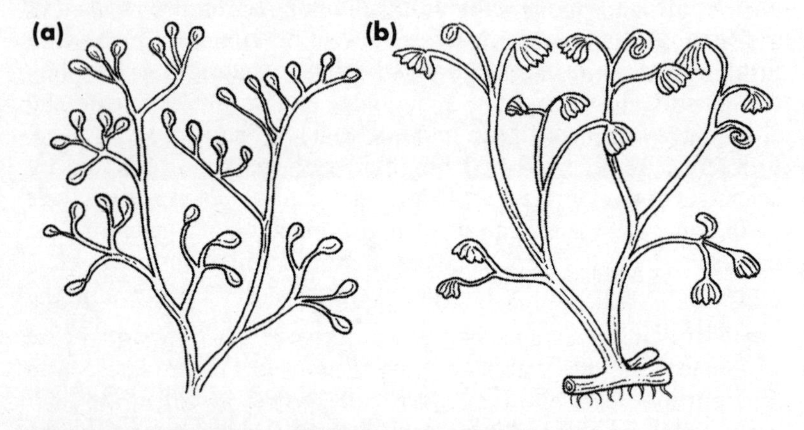

Figure 3.1 (a) Cooksonia; (b) Psilophyton

Soon a humus-rich soil was created. Plant stems grew increasingly taller towards the light and extensive **root systems** evolved to support aerial structures. By the early Carboniferous (365 million years ago) giant **clubmosses** and **horsetails** had reached heights of 40 m in their competition for energy-giving sunlight. Narrow leaves and cone-like spore-producing structures were some of their features and resembled those of modern conifers. As fossils, these plants have made a significant contribution to the coal measures of the world. In fact, they have played a part in the structure of buildings, too – in 1994, a builder knocking a hole in a wall at the Royal Museum of Scotland in Edinburgh discovered fossils of a 336-million-year-old swamp tree hidden in the stone! He was demolishing a wall of the museum when he noticed strange, dark lines running through the sandstone. The fossil was identified as a **lycopsid** or **scale tree** that grew in the Carboniferous period. The trees grew up to 35 m high with trunks 1 m across. They died out at the end of the Carboniferous when global warming dried up the swamps. These plants probably heralded the evolution of the conifers, but their reproductive methods remained primitive. Beneath the canopies of these giant primordial trees, **ferns** grew in abundance. With less access to light, these were the first land plants to develop full spreading leaves. Until the Lower Carboniferous period, all land plants were pteridophytes, which reproduced in alternate asexual and sexual generations. The **sexual generation** produced the equivalent of eggs and sperms from one type of spore. In more advanced types, however, such as the clubmosses, sexual reproduction entailed two different kinds of spores – the larger, female, **megaspores** and the smaller, male, **microspores**.

The sexual generation remained dependent on water for reproduction as a medium for the male sex cell to swim to the female sex cell. As a result, these early plants could not colonize any areas prone to prolonged water shortage. An evolutionary breakthrough came with the appearance of a group of plants called **cycads**. These resemble palm trees with large leafless stems and stiff leaves. About 120 million years ago they were widespread but today they are restricted to tropical and subtropical regions. They were some of the first of the higher seed-producing plants. A close living relative of the cycads has a good claim to being regarded as the most senior surviving plant in the world. It is the **maidenhair tree**, *Ginkgo biloba*, and is of the same generation as the dinosaurs, therefore

Figure 3.2 Leaves and fruit of *Ginkgo biloba*

being a 'living fossil'. Its peculiar leaves (Figure 3.2) are found with certainty in Permian deposits and probably in the Carboniferous.

Ginkgo biloba survived in Britain until the middle Tertiary. During the Jurassic period it reached its greatest abundance and almost world-wide distribution, probably aided by the prolonged resistance of its seeds to immersion in sea water, but it has declined progressively since that time and it is unlikely to be found in its truly wild state anywhere in the world. Until 1899 it was known only as a cultivated plant in China and Japan, where it has been grown from time immemorial as a temple tree. Indeed, it might have been saved from extinction only by its adoption by the Chinese as a sacred plant. The famous maidenhair tree of Kew Gardens in London was planted in 1760. *Ginkgo* is highly resistant to insects, disease and air pollution and is therefore an ideal 'street tree' in cities throughout the world.

The name *Ginkgo* is something of an etymological puzzle. It is thought to be an erroneous copy of 'Gin-ko', which is the Japanese equivalent of two Chinese symbols meaning 'silver fruit', a reference to its plum-like fruits. The seeds from its fruits are still sold in some Chinese markets as roasted *sal* nuts.

The Triassic and Jurassic periods (245–140 million years ago) saw the rise and fall of the giant horsetails and clubmosses, which were replaced by the precursors of conifers. During the Cretaceous period (144–65 million years ago) the land was transformed by the sudden success of the true **flowering plants**. The supercontinent Pangaea, which had been subtropical with dry deserts in its centre, began to drift apart at this time, causing changes in ocean levels and currents. The annual rings of fossil trees from the northern and southern extremities of the continent show distinct variations in growth rate at this time which indicate a change to a more temperate climate. As the land masses broke away and the Earth's crust stretched and compressed to form seas and mountain ranges, new habitats began to appear and a greater number of ecological niches became available for plants and animals alike. The flowering plants' adaptations to these changing environments proved extremely successful and within a relatively short time they dominated over their conifer-like ancestors and became the most widely distributed plants on Earth.

The spore-bearing parts and cones of conifers and the flowers of flowering plants are all types of modified leaves. The spores of most conifers compare with the pollen and eggs (**ovules**) of flowering plants. Also, the hollow tube-like cells for carrying water in woody flowering plants resemble similar cells found in conifers.

The sex life of flowers

It is important to consider the implications of the parallel evolution of several groups of marauding insects at the same time as flowering plant development. With this in mind, some suggest that in order to protect the naked spores of primitive plants from being eaten by insects, and as a protection against a desiccating atmosphere, the leaf structures which carried the female sex cell evolved to form an enclosed cavity, the carpel of modern flowers (Figure 3.3).

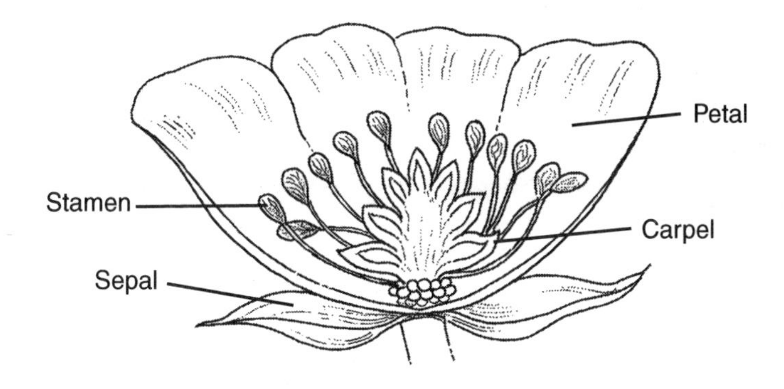

Figure 3.3 The structure of a typical flower

Protected in this way, the ovule could be a simple structure so that fertilization could be accelerated with minimal complication. Some fern-like plants, the **Bennettales**, in South Eastern Asia today, have their spore-bearing parts arranged in a way that resembles the rosette-like flowers of primitive flowering plants. Wherever and however they evolved, they had to follow similar strategies to the conifers in order to survive under the selective pressures of the environment. A major reason for their being today's dominant plants was their ability to exploit insects to help them reproduce.

The device they developed to use this allegiance was their characteristic reproductive structure – the **flower**.

The insects enabled plants' survival. In return the insects, often elaborately camouflaged, have used the very plants on which they feed and lay eggs as shelter. The plants have developed 'spare capacity' leaves and stems which can re-grow when eaten by pollen carrying insects. **Pollination** is the transfer of pollen from one flower to another, made possible by the parallel evolution of the flowering plants and insects. These relationships are beneficial to both plants and insects and remarkable adaptations have resulted by natural selection to accommodate each one.

Pollen-eating beetles were some of the first insects to be associated with flowers and as soon as this happened only those flowers that produced a superabundance of pollen were able to survive: some pollen would invariably be carried on the insects' outside rather than all of it being eaten. Some plants began to specialize in the production of sugary nectar to attract insects. By the beginning of the Cenozoic era (65 million years ago) beetles, flies, butterflies and moths had evolved from the basic ancestors of insects. They had developed special mouthparts which could reach down into the flower and suck up **nectar** from the **nectaries** which produced it.

Unlike the naked seeds of conifers, the seeds of flowering plants are enclosed within the ovaries of the flower (see Figure 3.3).When the mature seed is formed it consists of an embryo plant packed with food stores and wrapped in a protective coat (the **testa**). The protection helps the seed survive until conditions are suitable for germination.

Relatively little is known about the evolution of flowering plants. Most information has been pieced together from the comparison of living plants by arranging them from primitive types to more advanced. A fossil form found in Kansas was one of the only discovered members of an extinct family of flowering plants that lived about 95 million years ago. It was apparently a small tree or shrub living in the subtropical coastal plain among dinosaurs roaming the water's edge. Apart from this, fossil evidence of true flowering plants is scarce. A cynic once said that the Kansas fossil plant probably died out because of sheer boredom! Then in 1998 plant palaeontologists became more excited when a fossil flower, found in China, was dated as being 125–142 million years old. It did not seem to have had any conspicuous petals and until then, the record age for a fossil flowering plant (115 million years) was held by an Australian specimen. The Chinese find has been named ***Archaefructus*** (*Archae* = 'ancient'; *fructus* = 'fruit'). The fossil has the reproductive parts which define it as a flowering plant, together with the remains of mature fruits, 5 mm long with enclosed seeds. It is unclear whether it was pollinated by wind or insects.

Before this discovery, the flowers of primitive flowering plants were thought to have been large, with many stamens and carpels. This was based on the 95-million-year-old Kansas fossil magnolia-like flower. Even today similar types are quite common species. It demonstrates how primitive flowers first attracted insects. In modern plants the visiting insect, in its search for pollen for food, is attracted into the flower by scent and colour. It lands easily on the platform-like petals. While it crawls around the centre of the flower collecting food, the rough sticky pollen grains from the stamens become caught on the insect's body. Flying to the next flower, it involuntarily deposits the pollen on the sticky stigmas of the carpels.

Many plants, however, rely on wind for pollination. Their flowers are small and inconspicuous. They have stamens positioned outside the flower so that the pollen can easily be blown away. Their stigmas are often feathery to act like a net in catching the pollen on the wind. Pollen must be produced in astronomical abundance for this 'chance' method of pollination to be successful – and hay-fever sufferers become seasonally and distressfully aware of this fact.

Some plants have evolved truly astounding pollinating mechanisms to ensure survival. Perhaps the most astonishing ones are those that lure insects by disguise. The carrion flower, *Stapelia*, for example, produces such a powerful smell of decaying flesh that flies cannot resist it and lay their eggs in its petals. When they fly to another flower, they carry some pollen with them. Another example is the bee orchid, *Ophrys apifera*, which has a petal that resembles a bee. In an attempt to fight or mate with it, a male bee inadvertently pollinates it.

Members of fifty major groups of flowering plants have been found as fossils dating back to 100 million years ago. By the end of the Cretaceous period, all the subdivisions of the plant kingdom were present. The 23 000 or so species distributed across the world today provide the first link in food chains for many animals. Without plants in the first wave of the invasion of land, evolution of the diversity of terrestrial animals would not have been possible.

The beachhead established

Vertebrates established themselves on land during the Devonian. Ironically, the invasion of the land is thought to have come about to enable the animals to sustain their lives in water! As we have seen, the first wave of the invasion heralded the success of land plants and some arthropods, so there was plenty of potential food already on land for the first amphibians to exploit.

The plants consisted of ferns, horsetails and tree-like clubmosses. Primitive insects and spiders made a living among the lush swampy vegetation. The Devonian was a period of great climatic and geographical changes, particularly in the northern hemisphere. The continents of North America and Europe had moved together, obliterating the intervening ocean and building a vast mountain range. However, these mountains soon began to erode and only the vestiges can now be seen in the Appalachians, the Scottish highlands, and the mountains of Norway. Their exposed rocks were broken down to vast quantities of sand which eventually became Old Red Sandstone. The shallow lakes, rivers and estuaries became the main home of the first fish-like vertebrates. These waters were potentially hazardous environments because they were subject to so much change. Long periods of hot dry weather would have dried up bodies of water leaving small stagnant pools containing little oxygen. Only those creatures that could obtain oxygen from the air as well as from water would have been able to survive. Most of these early shallow-water fish developed a primitive lung-like breathing surface that enabled them to gulp air from the surface of the water and extract oxygen from it. The so-called bony fish (those with skeletons made of bone rather than cartilage) that later returned to the sea were able to use this lung as an organ to adjust their buoyancy. We find it in today's fish as the **swim bladder**.

Under these progressively arid conditions, the pools would dry out completely. Only creatures that could burrow into the mud until the rains came, or those that could scramble over land to the next habitable pool, would survive to breed and pass on their genes that controlled these adaptations.

A fish swims by moving its body and tail in S-shaped curves but its streamlined shape and such movements would not help it to move over land. During the Devonian, however, fish developed a type of skeleton to support their fins that was to prove essential in the conquest of land. The paired fins of fish stabilize them in water, and when strengthened they become more efficient. In one group they became lobe shaped with a fleshy base supporting a fringe of fin rays. The lobed fins were the forerunners of legs of land-dwelling animals.

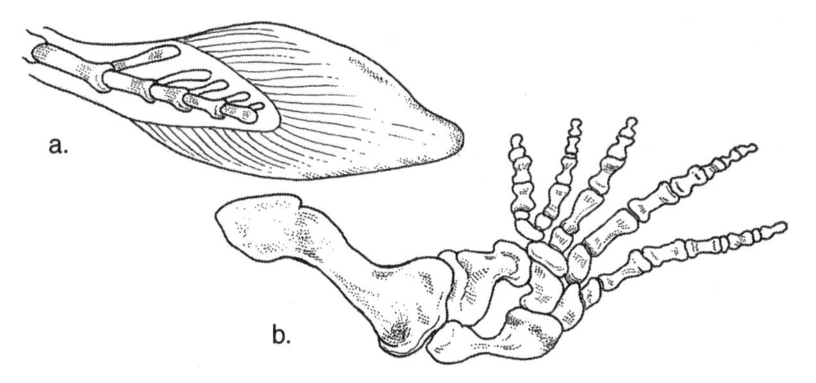

Figure 3.4 Limb development. (a) The bones of the lobed pectoral fin of ancient fish developed into the bones of the forelimbs of the land-living vertebrates. (b) The limb bones of an amphibian can be called 'pentadactyl' or 'five-fingered'. Note that most of the small 'fish bones' have been lost and the remaining ones are longer and stronger. This first amphibian-style limb was the forerunner of the limbs of all land-living vertebrates

Lobed fins provided the support to prop up those fish that were stranded in drying up pools, enabling them to lever themselves forward by using diagonally positioned limbs while throwing their bodies into the familiar fish-like movements of S-shaped bends. Such fish became caught in stagnant and drying inland seas or streams, and were able to survive by breathing while they dragged themselves on their stumpy little lobes to better water. Without these 'limbs' they would have perished entirely, as doubtless other water life around them did. Perhaps they were persuaded to loiter in the mud by such tasty and easy prey; perhaps at first they were lured by fresh bodies of water, and later simply by fresh bodies!

It is likely that **amphibians** arose from several groups of lobed finned ancestors similar to the coelacanth (see page 74) because several types of Devonian lobe-fins have skeletal features that resemble those of later amphibians. Some had two pairs of lobed fins that were well placed for movement on land and others had nostrils similar to those of frogs.

A fossilized skeleton of the earliest true amphibian was found in the 1930s in the Upper Devonian deposits in Greenland and many major new discoveries were made in the 1980s and 1990s. It is called **Ichthyostega** and is one of the most perfect fossil links between major groups that has ever been found. However, unmistakable amphibian footprints have been found in earlier Devonian deposits in Australia. Ichthyostega has a combination of fish and amphibian features. Its sense organs and long-finned tail are more fish-like characteristics, whereas its sprawling limbs and expanded ribs are more like those of amphibians (Figure 3.5).

Figure 3.5 Ichthyostega

Although Ichthyostega is the oldest complete amphibian skeleton yet known, it is unlikely to have been the ancestor of today's amphibians as it was probably one of the many amphibian 'experiments' of the Devonian, most of which have ended as fossils with no evidence of surviving direct descendants.

It was during the Carboniferous period (360–286 million years ago) that the amphibians flourished and spread. Many early types were much larger than those existing today – indeed, the Carboniferous era is often called the **Age of the Amphibians**.

The harsh desert-like conditions of the Devonian acted as an environmental selective pressure which forced restrictions on the basic shape of amphibians. They typically developed the four-limbed stance with the head above the ground and a barrel-shaped body holding the lungs. Climate changes and the consequent lush fertile swamps of the succeeding Carboniferous provided a variety of habitats eminently suitable for the amphibian mode of life. As a result the amphibious group spread by adaptive radiation and colonized the warm shallow waters of marshes as well as returning to a totally aquatic lifestyle – as we see in newts and some salamanders today.

From very early times the amphibian group split into two main evolutionary lines, which were united by the fact that they probably laid eggs in water and their young went through a tadpole-like larval stage – just as modern forms do. The fossils of one group reveal skeletons that suggest they gave rise to **reptiles**, whereas the other group may have been the ancestors of today's amphibians.

The group that gave rise to reptiles had an arrangement of limb bones which became the basic pattern of the limbs of all land living vertebrates (the **pentadactyl limb**, see Figure 3.4). After the Carboniferous came the Permian (286–245 million years ago), when arid conditions returned to the northern continents. Pressures were again put on aquatic animals. Those that could live for longer periods out of water had an advantage over those that could not. Some predominantly land-living amphibians had developed into many different forms. Fossils of types with skulls more than 1 m long have been found and even 5 m long sea-living forms existed in the Upper Carboniferous period.

One type showed advanced skeletal features of the reptiles: the hip bones were more securely joined to the backbone and the number of bones in the toes increased. Both of these factors enhanced movement on land. These creatures were considered as the first reptiles but their skulls remained primitive and there were still some sense organs that resembled those of fish, showing that they were still essentially amphibians.

Modern amphibians are poor relatives of the spectacular creatures that existed in the Carboniferous and Permian periods. They breathe mainly through their skin because their lungs are very simple and their rib cages are not strong enough for breathing movements normally associated with totally air-breathing vertebrates. Extinct forms had strong rib cages and skin that was often covered with armoured plates. In the Permian and the Triassic (245–202 million years ago) amphibians, with their dependence on water for laying eggs and for their tadpole larvae, were no match for the more efficient reptiles and so they declined. You don't see many amphibians today. The only survivors are about 1500 species of water-dependent forms. They have become so specialized that they have lost many of the advanced features that made their ancestors so versatile and successful 400 million years ago. The largest living type is the giant salamander of Japan which grows to about 1 m long. A fossil of this type has been found in Miocene rocks (24 million years old) in Germany. A specimen found in 1726 was originally thought to be the remains of a sinner drowned in the biblical flood! The reference, of course, was to the deluge of the time of Noah which allegedly took place four or five thousand years ago. It was named *Homo diluvii testis* – the ' man who witnessed the flood'.

The long march inland

The amphibians got us part way out of the water, and we should not sneer at their somewhat clumsy gait on land. After all, they added true pentadactyl limbs to the fish torso instead of the stumpy lobes of the lobed finned fishes. They also gave us a model for our ears: our eardrums have developed from membranes from the sides of their heads and the tiny bones of our middle ears have been borrowed from their jaw supports. Unfortunately for them, however, the amphibians' debits eventually outweighed their credits:

- Their eggs are small, like those of fishes, and must be protected from drying up by being laid in or near water.
- Their tiny larval stages, like tadpoles, are also very vulnerable and must live in water because they use gills to breathe.
- Many develop lungs in the adult stage but these are invariably inefficient, as are their arrangements for blood circulation.

In short, in a real invasion of land the amphibious troops would be considered to be a pretty makeshift outfit with little chance of success and would be replaced by more robust troops with proper equipment to overcome the rigours of establishing themselves on land.

By the time the extensive swamps of the Carboniferous had dried out to make way for the bleak deserts of the Permian and Triassic, the reptiles had become total masters of the land and were to remain so for the next 200 million years. The evolutionary leap between the ancestors of amphibians and the ancestors of reptiles was possible only because of the development of a special type of egg that could be laid on land (the **cleiodoic** or **amniotic** egg).

An amphibian's egg is laid in water with little protection. In contrast, a reptile's egg is laid on land and has a leathery shell that allows for the passage of oxygen in and carbon dioxide out. It is a complex structure containing the embryo, a large store of food (**yolk**), a system for absorbing waste products (the **allantois**), and an internal membrane (the **amnion**) that envelops the **embryo** in its own little private pond. This type of egg is a self-supporting capsule that enables survival in the hostile environment of dry land and was a forerunner of a bird's egg. So the egg really did come before the chicken!

In the Eden of the amphibians, the Carboniferous swamps were ideal for many types of creatures including predators that would readily eat unprotected eggs and larvae. Today, some amphibians do not lay eggs in open water (many tropical frogs lay them in water-filled hollows in trees; some even carry them on their backs or in throat pouches for protection). Predation became a selective pressure behind the development of the reptilian egg that could be laid (and possibly buried) on land.

At first reptiles lived in water like amphibians and, before they became adapted to a totally land-dwelling existence, their backbones became stronger and less flexible and their hip girdles became fixed to the backbone by two vertebrae rather than by one (as in amphibians). The number of bones in the toes increased, giving their feet a better grip on the ground. In order to attack and hold prey on land, an animal needs to have more powerful jaws than it would need for feeding under water. The consequent muscle development required modifications of the skull bones and gaps between the bones to make room for their attachment. The number and position of these gaps formed the basis for the classification of fossil reptiles (Figure 3.6).

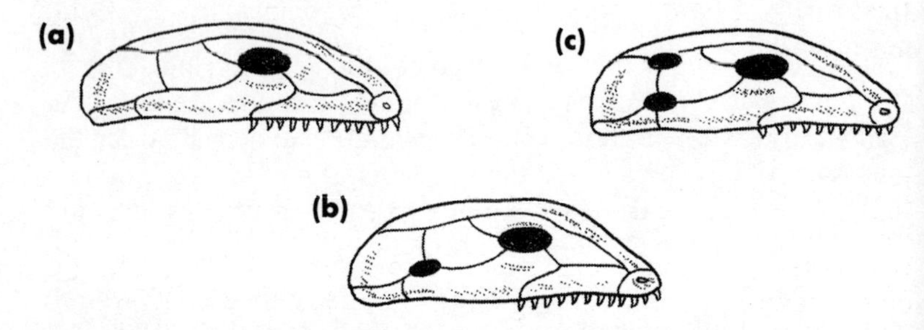

Figure 3.6 Skull patterns in reptiles: (a) anapsid; (b) diapsid; (c) synapsid

The most primitive type of reptilian skull has no gaps (**anapsids**). These so-called 'stem' reptiles, the **cotylosaurs**, have modern equivalents in the tortoises and turtles which evolved in the Triassic period. Those types of skulls with a single gap low down on each side (**synapsids**) were probably the forerunners of the mammals. **Diapsids** had two gaps on both sides of the skull. They gave rise to a group called the **archosaurs**, which were very significant in evolutionary terms because they split into four lines, two of which were to become the dinosaurs.

The remainder of the reptilian ancestors had a gap high up on each side of their skulls (**euryapsids**). They included the now extinct great swimming reptiles, the **ichthyosaurs** and **plesiosaurs**.

By the end of the Triassic, at least one group of these early reptiles died out, leaving their diminutive descendants – the **mammals**. The archosaurs were then able to leave the water and live in relative safety. Out of water the long tails and hind legs of these animals gave them their characteristic stance. They walked on their hind legs, pivoted at the hips, with the weight of their bodies counterbalanced by their tails. At this stage the archosaurs diverged into four different directions. Some, the **crocodiles**, stayed in the same habitat and their descendants have hardly changed to this day. A second group took to the skies and became the **pterosaurs**, leaving the other two groups to become the **dinosaurs**. Surely an alien from space visiting our planet at that time would be forgiven for thinking that the vast diversity of reptilian fauna represented the final rulers of Earth's animal kingdom, from whose grip it would never escape.

Enter the dragon

Despite the fact that no human has ever seen a living one, perhaps no other extinct group of animals generates more interest than the dinosaurs. The interest knows no age barrier. From 5-year-olds to octogenarians their enduring popularity has made them the entrepreneur's 'golden egg-laying goose'. Apart from some of the best money-making films of all time, toy shops are full of them, television series are devoted to them and small ones have been known to fall out of packets of breakfast cereals! Their popularity does not stop at trivia though; scientists are also fascinated by these extinct wonders of the ancient world. Palaeontologists spend years debating their lifestyle. Were they warm blooded? What colour were they? Could the largest of them run? Why did they suddenly become extinct? There are more than a thousand books in print in English alone which feature dinosaurs. So what was so special about the dinosaurs as an evolutionary experiment?

Fossils of dinosaurs have been exposing themselves in eroding rocks for millions of years all over the world. The ancient Chinese believed that they were the bones of dragons and treasured them because of their reputed medicinal and magical properties. Probably the earliest written record of a fossil dinosaur is a Chinese description of a dragon bone, made over 1700 years ago. Early American Indians sometimes found fossil bones and teeth which they thought were from giant buffaloes. The young braves kept them as good luck charms on their hunting expeditions.

An interesting interpretation of what turned out to be part of a dinosaur's thigh bone was made by R. Brookes in 1763 (Figure 3.7) He believed it to be a scrotum of a giant human and named it *Scrotum humanum*! In fact, this fossil had been found many years before – in 1676 – and is probably the first recorded dinosaur find in Britain. Robert Plot, the Keeper of the Ashmolean Museum in Oxford, was sent the object from a local quarry. He published an illustration of it and suggested that it was the knee end of a thigh bone of some giant animal, possibly from an elephant brought over to England by the Romans. There is no evidence to show that Romans ever brought elephants to Britain and so he changed his mind and said it was from a man or woman '... notwithstanding their extravagant magnitude'.

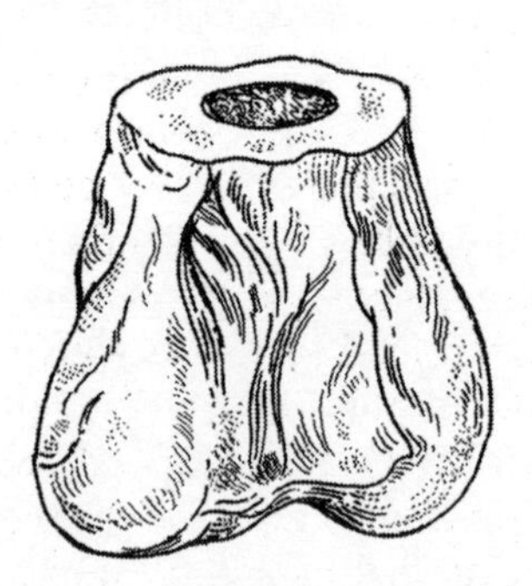

Figure 3.7 *Scrotum humanum* of the seventeenth century

Fossil footprints have been found in abundance throughout the nineteenth and twentieth centuries but were at first thought to be those of giant birds. With all this accumulated evidence of fossilized former life-forms, why were scientists of the eighteenth and nineteenth centuries so reluctant to recognize them for what they were? The religious dogma which saturated Europe and America at the time had precluded any thoughts other than 'God had made the Garden of Eden with all the creatures that have ever lived'. Why should God go to the trouble of creating beasts and then let them die out?

It was not until Lyell's classic work on geology (see page 11) that hard evidence started to emerge to show that the Earth's rocks were hundreds of millions of years older than the biblical creation date of 4004 BC suggested. The idea of fossils not being placed in the Earth by the devil to fool us was controversial enough, but to propose the idea that there once had been giant reptiles on Earth millions of years before humans was simply unacceptable to many people. The pioneer collectors of dinosaur fossils had much academic courage and brilliant deductive skills to challenge current views on the prehistory of Earth.

One of the first attempts to bring dinosaur fossils to the notice of the scientific world was made by **Gideon Mantell**, a physician from Lewes, set in the fossil-rich countryside of Sussex. Here he practised amateur palaeontology and made a large collection of fossils. Legend has it that in the spring of 1822, Mantell and his wife Mary Ann visited a patient and while he was tending to the patient his wife went for a stroll along a country road where workmen were carrying out repairs. She noticed something shining in a pile of rocks which had been dumped there from a quarry for use in the road mending. On examination she saw they were some strange-looking fossilized teeth. Her chance find was one of the most important events in the whole history of the study of dinosaurs. She gave them to her husband, who had not seen anything quite like them before but thought that they must have come from a very large animal indeed. Within his experience, the only things that he could compare them with were teeth of an elephant or rhinoceros. He made enquires about the source of the stone and these led him to a quarry in the Tilgate Forest. The age of the rocks from this quarry was about 130 million years – the Mesozoic Era. No mammals lived in those days, so he deduced their pre-mammalian origin and suggested that they were from a giant prehistoric reptile.

The expertise of a comparative anatomist was needed to ascertain what sort of 'monsters' had left such remains and so Mantell sought the opinions of Dr **William Buckland**, Professor of Geology at Oxford University, and the eminent French anatomist Baron **Georges Cuvier** (1769–1832), who had been one of the first scientists to use fossil evidence to demonstrate that extinction had happened. Cuvier's studies on fossil bones quarried from the Paris Basin established the former existence of exotic animals. These fossils, and others, including giant land animals from America, effectively suggested the idea of extinction. Cuvier visited England in 1818 and examined many of the early collections, but he died before he could publish his conclusions.

The responses to Mantell's find were far from encouraging. Cuvier thought that the teeth were from large mammals and Buckland was of a similar opinion or thought that they might be from a large fish. Buckland gave a friendly warning to Mantell not to publish his description! Despite this, Mantell began searching through collections of ancient and modern skeletons for anything that might give him a clue to the animal to which his fossils belonged. At last, while examining a collection at the Hunterian Museum of the Royal College of Surgeons, he found what he was looking for. A jaw of an iguana from South America had mini-versions of the fossil teeth that his wife had found three years previously. Mantell published his description and named the animal **iguanodon** ('iguana tooth').

There was another lapse of time before further progress could be made. Then, nine years after Mantell's publication of the iguanodon find, a slab of rock from a Maidstone quarry was discovered crammed with what turned out to be iguanodon fossil bones. Now started what was probably the beginning of the commercialization of dinosaurs ,which was to continue to Spielberg's film *Jurassic Park* – and which will probably continue until humans become extinct. The quarry owner at Maidstone recognized a fine business opportunity and offered the fossil-filled slab to Mantell for the then ridiculously high price of £25. Mantell's friends clubbed together to buy it for him and it was delivered to his home in 1834. His family put up with three months of his chiselling away the rock from the precious remains. His claim to fame is recognized by the fact that the town of Maidstone has a coat of arms with an iguanodon on it.

After the scientific acceptance of the nature of iguanodon, Buckland had egg on his face, having discouraged Mantell's publication. However, he too had an input into early dinosaur discoveries – in 1824, a year before Mantell's official published account of iguanodon, Buckland had described the jaw of an animal, which he classed as **Megalosaurus**. William Buckland became Dean of Westminster Abbey but is probably better known as an eccentric with an interest in geology. His eccentricity was demonstrated by his frequent visits to the local post office with his pet bear, which used to raid the sweet counter! He also kept jackals in his living room. Despite the published descriptions by Buckland and Mantell and the descriptions of seven other Mesozoic reptiles by 1840, the term 'dinosaur' had yet to be invented.

The terrible lizards

The two dinosaur groups, **lizard-hipped** and **bird-hipped**, were no more closely related to each other than they were to the crocodiles or the pterosaurs, so the term 'dinosaur' really has no scientific meaning in the context of classification. The name comes from the Greek words *deinos* (meaning 'terrible') and *sauros* (meaning 'lizard') and was first coined in 1842 by Sir **Richard Owen**. Although an opponent of Charles Darwin's theory of evolution because he resisted the idea of humans and primates having a common ancestor, Owen was one of the pioneers of vertebrate palaeontology. He became an expert anatomist, gaining his skill by dissecting animals that had died at London Zoo. Subsequently he applied his skills to the difficult problem of reconstructing extinct forms from fragmentary fossil material. The reconstructions from fossil bones that Owen made often resembled gigantic versions of known reptiles so his name for them is not entirely unexpected. Owen's interest arose from a collection of fossils made by **Mary Anning** (1799–1847), the first professional fossil collector and sales woman and the celebrated 'seller of sea-shells on the sea-shore'. Between 1824 and 1828 Anning collected the first complete skeletons of ichthyosaurs, plesiosaurs and pterosaurs along the Dorset coast. She sold many of her finds to tourists in her father's shop in Lyme Regis. In this part of the world, souvenir shops still sell locally found fossils, and it is known as 'Europe's premier dinosaur locality'. The rocks of the area date from the Cretaceous (up to 140 million years ago) when the region was a flat coastal plain inhabited by ancient reptiles. Since before recorded time, winter storms have battered the coastline, frequently eroding a couple of metres from its unstable rocks and uncovering the burial ground of the lost giants of old. Fossil trees, dinosaur footprints and a variety of reptilian fossils are revealed each winter to professional and amateur palaeontologists alike. A few fossils remain for summer tourists but not all fossil searchers are lucky enough to find them. Reputed 'fossilized teeth from meat-eating dinosaurs' are often taken to the curator of the local museum by eager schoolchildren but almost invariably these 'fossils' turn out to be tooth-shaped flint nodules.

Owen could have had no idea of the show-business potential of his invention of the name 'dinosaur' in 1842. On New Year's Eve 1853 he and 21 other scientists of repute dined inside a massive model of a dinosaur in the grounds of the Crystal Palace, London.

After the death of Cuvier a power struggle developed among English anatomists to inherit Cuvier's distinguished status as the best comparative anatomist in Europe. It included the amateur, Mantell, and the professionals, Owen and **Robert Grant**, who was Professor of Comparative Anatomy and Zoology at London's University College (known as the 'Godless college'). Grant lived up to his institution's reputation by believing that evolution was an explanation of the fossil record. Dinosaurs became the weapons in this battle and the battle lines were drawn up between religious beliefs and theories of evolution – remember, Charles Darwin had yet to publish *The Origin of Species*.

Owen was a rising star at the Royal College of Surgeons and was championed by the British Association for the Advancement of Science. Mantell was supported by the Royal Society but Grant had to make do with what he could earn from teaching. Owen saw dinosaurs as four-footed reptiles, but similar to present-day elephants. Such reptiles seemed to him to disprove the arguments of those who favoured progressive evolution, because God had clearly given these extinct reptiles the characteristics of advanced mammals at their creation. Owen took dinosaurs as a weapon in his battle with Grant, who believed that evolution explained variations in fossil form. The naming of a new order of reptiles sounds a most unlikely money spinner, even though Owen scored highly against his two main scientific rivals.

By the mid 1800s, the quest for dinosaur remains had spread to the USA. A chance conversation between **William Parker Foulke** and a local farmer in the town of Haddonfield in 1858 led to Foulke searching a marl pit on the farmer's land. Twenty years earlier, locals had dug up some massive fossil bones in the same area. A large number of fossil bones were found and were identified by Dr **Joseph Leidy** of the University of Philadelphia as those of dinosaurs similar to iguanodon.

The next major dinosaur discovery took place in April 1878, when some miners in Bernissart, Belgium, found themselves digging through a giant fossil some 322 m below ground. They were in a 'mass grave' containing the skeletons of 30 iguanodons, which probably met their death by falling into a deep, narrow ravine and were buried over a period of years by sediments deposited by a river. Hundreds of fossil bones were recovered from the mine, numbered block by block, and transported to Brussels. The task of piecing the giant jigsaw together fell on the shoulders of the newly appointed assistant naturalist of the Royal Museum of Natural History in Brussels, **Louis Dollo**, the first person ever to have the opportunity of building a whole dinosaur from its original bits and pieces like a giant construction kit. The only building big enough to house the titanic iguanodon was a church in Bernissart, and there the task was completed. However, most of these skeletons can still be seen in vast glass enclosures in the Brussels museum.

At the same time that Dollo was assembling his giants in a church, the great *American Dinosaur Rush* was taking place. Two distinguished academics, **Edward Drinker Cope** and **Othniel Charles Marsh**, instigated a widespread search in the USA, which resulted in the discovery of almost 130 dinosaurs new to science. The two palaeontologists were rivals to the extreme. There is a story which tells of Cope's reconstruction of a fossil marine reptile being criticized by Marsh because it was back to front, with its skull at the end of its tail. Cope never forgave him for this public humiliation.

In the late 1880s unmapped parts of North America were not the safest of places for academics to wander around. In 1876 when Chief Sitting Bull was busy chopping up General Custer's soldiers at Little Big Horn, Cope was leading an expedition to Montana, much against advice from people more knowledgeable about that area. However, he survived and one of his team is quoted as saying that his close encounters with Indians were friendly, with Cope entertaining the locals by taking out his false teeth and showing them to the natives! Whatever the truth is about the rivalry between Cope and Marsh, these two intrepid dinosaur hunters will be remembered as men who revealed the world of dinosaurs as a varied community of strange reptiles.

By the early part of the twentieth century fossil dinosaurs had been found on every continent. In 1907 a German engineer noticed gigantic fossil bones just outside the village of Tendaguru, Tanzania. He reported the find to the director of the Berlin Museum and thus began the biggest dinosaur excavation ever undertaken to that date. Over 500 local workers and their families camped at the site which was four days march through the jungle from the coast. It became a little town that had to be provided with water, food, sanitation, building materials and medicines. In the first three years, 4300 loads of fossils were sent to the port of Lindi – a total of 250 tonnes.

Dinosaur discoveries in Mongolia were quite accidental, insofar as they were found by an expedition looking for early human fossils. The result was the excavation of the first fossil dinosaur nests with eggs. Between 1922 and 1925, a series of expeditions to the same place revealed many new and spectacular dinosaurs. Between 1933 and the 1990s Chinese palaeontologists have introduced to the world a fascinating collection of dinosaurs from virtually every known group.

Although knowledge of the two main branches of dinosaurs increases year by year, it is still quite limited. Most is known about the 'bird-hipped' (**ornithischian**) group (Figure 3.8). They had well armoured heads and their massive skulls have been found in river and swamp sediments where they were carried by mountain rivers and streams. In contrast, the skulls of swamp-dwelling types are usually poorly preserved because they were usually lightly built and easily crushed beyond recognition.

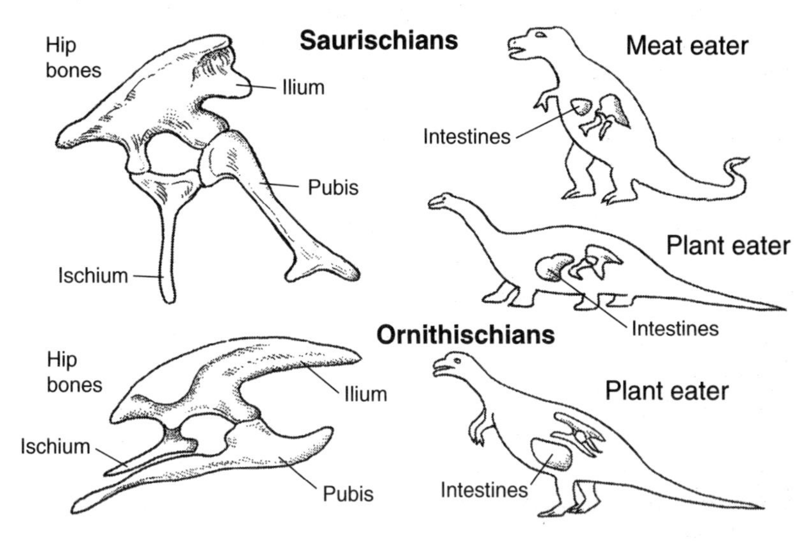

Figure 3.8 The two main types of dinosaurs

The other main group, the 'lizard-hipped' (**saurischian**) types (Figure 3.8), appear in earlier deposits than the 'bird-hipped' dinosaurs. They looked like the bipedal crocodiles that immediately preceded them and were all carnivores, therefore hunting the contemporary herbivores. The first dinosaur ever to be scientifically described by Buckland in 1824 was the saurischian called **Megalosaurus**. It was a 7 m long creature with three huge claws on its hind feet. Its skull was equipped with serrated meat-shearing teeth and its head may have been held up on an S-shaped curve of its short neck, like that of a bird.

In the Cretaceous period the monster-like saurischians developed in many different ways, some with huge spines on their backs and others, like the mighty **Tyrannosaurus**, with 20 cm long teeth.

Despite this animal's legendary reputation as a titanic killing machine, it seems likely that it was not a fierce predator but lived as a scavenger, feeding on the corpses of other animals and leftover meals of more active, if smaller, hunters. It is possible that these hunting dinosaurs had a constant body temperature, which was a pre-requisite to giving rise to their descendants – the birds.

The heavyweight giants of the dinosaurs appeared in the Triassic. For example, **Diplodocus** was 25 m long and probably weighed some 30 tonnes. They lived in the lush swamps in the Jurassic and Cretaceous eras, feeding on the abundance of marshland plants.

Like the saurischians, the ornithischians also developed a four-footed stance and became the 'armoured troops' of the land invasion. Some resembled the modern-day rhinoceros, with horns and bony frills protecting their necks. Many different types of armour and defence mechanisms were developed. There were types with spikes that ran down their bodies and on to their tails. They used these to ward off aggressors. It has been suggested that the plates running down the backs of some dinosaurs were used for heat regulation – like radiators – rather than for defence. These plates could not have been very strong because they seemed to be merely embedded in the skin and not fixed to the skeleton.

Rapid advances

Between the Triassic (245–202 million years ago) and the Cretaceous (140–65 million years ago), some reptiles abandoned their land-living way of life and reverted to the aquatic lifestyles of their ancestors. One selective pressure at this time was the spread of inland seas and the flooded continental shelves of the Pangaea supercontinent. Fish had by this time spread through the seas from their inshore habitats of the late Palaeozoic, providing an abundant food supply. Other land animals became airborne. In their case, the adaptations necessary for flight ensured escape from predators rather than to obtain food.

Perhaps the most notorious of the marine reptiles were the fish-like **ichthyosaurs**. They had a dorsal fin like a shark and limbs that were modified as balancing and steering paddles. Their tails were finned and could propel them rapidly through the water. They first appeared in the Triassic and there is an abundance of their fossils in marine deposits. Unlike the reptiles of today, the ichthyosaurs were unable to lay eggs on land and so must have given birth to live young in water, having retained the eggs until they hatched inside them. Competition from other marine reptiles led to their extinction at the end of the Mesozoic era.

The other large and well known group of swimming reptiles of the same period were the **plesiosaurs**. Two main lines of these animals developed from a common ancestor in Triassic times, both very different. Their ancestors were shore-dwelling fish eaters with long multi-toothed jaws, long necks, webbed feet, and a fin along the top of their long tails. Their descendants were far more specialized for life in the sea but not as well adapted as the ichthyosaurs. One group had squat bodies but extremely long necks. Their limbs had developed into paddles and drove them through the water with a 'flying' action rather like that used by penguins today. Steering was accomplished with some help from a diamond-shaped paddle on the end of a short tail. They lived by catching fish near the ocean's surface, using their long necks to help catch prey. The other group had short necks and were much larger – some 12 m long fossils have been found in the Cretaceous deposits of Australia. They had very large heads and streamlined bodies and must have represented the reptilian equivalent of today's sperm whales, also fishing for giant squid in the ocean depths.

The three major groups of marine reptiles probably did not compete with one another for food but hunted different prey at different depths in the oceans.

Until the Triassic, the only land invaders to become airborne were the insects. **Gliding reptiles** appeared in the Triassic period when a number of lizard-like animals developed membranes supported by extended ribs. It is likely that those with such adaptations were at an advantage in escaping the small carnivorous lizard-hipped dinosaurs of the same period.

The reptilian masters of the sky were the **pterosaurs**, which, like the dinosaurs, were descended from the archosaurs. Their flying ancestry can be traced back via a 30 cm gliding reptile from the

Triassic deposits of central Asia, **Podopteryx**. This creature had a gliding membrane attached to elongated hind limbs and tail.

The best known examples of pterosaurs arc small pigeon-sized reptiles that flew by means of a pair of leathery wings which were attached to their bodies, hind legs and arms. They could fly relatively efficiently, altering the wing area by stretching and folding their fore- and hind-limbs to manoeuvre. They probably were fish eaters, snatching them from the water surface.

To enable the pterosaurs to sustain their active lifestyle they were warm blooded, and traces of hair or fur have been found among their fossils. The best known large pterosaur is **Pteranodon**, found in the USA in late Cretaceous limestone in 1872. It had a wingspan of 6 m. In 1975 a gigantic species, **Quetzalcoatlus**, was found in Texas, with an estimated wingspan of 11–15 m – probably the largest flying animal ever to have soared through the skies of our planet. It would have been at least three times as big as any flying bird and was more like an aircraft than any animal living before or since.

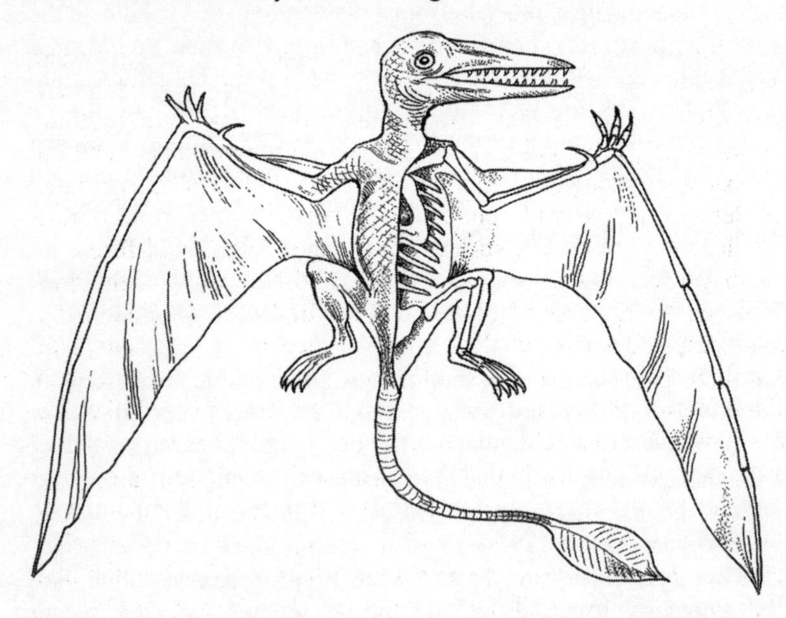

Figure 3.9 A pterosaur

The decline of the age of the reptiles came at the close of the Cretaceous about 65 million years ago (there is more about the possible reasons for this on page 184). The dinosaurs died out, as did the great sea-dwelling reptiles. Many invertebrate groups became extinct at the same time.

There are two main school of thought concerning this mass extinction:

- The first theory is that some sudden catastrophic event wiped out the big reptiles and the marine invertebrates, but allowed the smaller reptiles, the mammals and the birds to survive.
- The second theory is that some gradual factors, involving climatic changes, alterations of the geography and slow changes in vegetation, were the cause. Although such changes would have been gradual they may appear sudden in the geological record, where a million years is almost a blink of an eyelid in relation to the age of the Earth.

The supercontinent of Pangaea that had been in existence throughout the Triassic and Cretaceous had began to break apart and individual continental pieces were drifting away. The resulting loss of extensive inland and continental margin seas, coupled with the extensive mountain building, would have changed the climate. Evidence for this is the change of vegetation at the end of the Cretaceous period. The sudden appearance of tropical forest in North America showed a rapid warming of the climate, and this was followed by the spread of temperate woodland, dominated by conifers as the climate cooled. Animals, such as the dinosaurs, that had evolved to live under stable conditions during the previous hundred million years, could not have tolerated changes like this and subsequently would have perished.

With the extinction of the dinosaurs and other large dominant reptiles, the mammals and the birds spread and occupied the niches that had been vacated. Within a few million years the forests that had been browsed, seas that had been fished, and insects that had been preyed on by reptiles were being browsed, fished and hunted primarily by mammals and birds.

The reptiles that survived can be divided into three major groups:

- The tortoises and turtles
- The crocodilcs and alligators
- The lizards and snakes

There is also one surviving representative of an ancient group, the **Rhynchocephalia**. This is the **Tuatara** or **sphenodon lizard**, which now only survives on a few protected islands near New Zealand.

The airborne divisions

Two totally unrelated groups of animals, the insects and birds, have dominated the air during the course of recent evolution. Apart from flying reptiles which ruled for about 200 million years, no other type of animal has made significant inroads to this mode of life. Even the mammalian bats are comparative newcomers to an aerial existence, with the earliest fossils, found in North America, dating from the Eocene (58 million years ago).

The first insects did not have wings. Their fossils can be seen in the early Devonian sediments in Scotland. The evolution of the more complex land plants may well have coincided with the major colonization of the land by arthropods because those plants would have supported a great variety of animal life. The next insect fossils that were found are more than 50 million years younger – from the Carboniferous. This gap illustrates a general rule of palaeontology; that organisms living on land are less likely to become fossilized than are marine creatures. This is because most sediments which later turn into fossil-bearing rocks originally accumulate below sea level. By the time of the Carboniferous there was a greater variety of land animals and plants for these arthropods to feed on.

The Carboniferous insects were much more advanced than the first wingless types (springtails) and included a variety of winged insects. Most of them belonged to now-extinct groups, of which the biggest were the giant dragonflies, with wingspans up to 0.7 m – much larger than any insects living today. The reason for this gigantism has been a point of speculation, but one fact that emerges from the fossil record is that insects probably had no competition in the air at this time, there being no birds or any other kinds of aerial predators. A few of the Carboniferous insects are still familiar today, the most abundant of which are the cockroaches.

The first appearance of types of insects that have larval stages occurred in the Permian period (286–245 million years ago), at a time when many other groups of organisms were becoming extinct. These insects included the beetles, butterflies, moths, and bees. The young of these types spend most of their time feeding, whereas the adults are concerned mainly with reproduction. The reasons for the success of insects are probably very diverse and complex, but one important factor in their favour is their short life cycle and their ability to adapt to changing conditions; another is their early evolutionary development of the power of flight, perhaps initially to escape from predators. Once in the air, insects could have populated places that ground-dwelling animals would have found difficult or impossible to reach. This may have given them a particular advantage over other early invertebrates.

One flew over the dinosaur's nest

Eventually, the vertebrates evolved types that took to the air. Today the dominant types in terms of numbers and diversity are the **birds**. The size and strangeness of the dinosaurs often makes us forget that they lived in a time of great diversity of life. Even in the early Mesozoic primitive trees and conifers flourished, offering habitats to many small creatures. Insects were abundant, but other animals were unable to fly and as a result could not exploit the full potential of the forests. They may have been able to climb and jump from branch to branch, but at the ends of the twigs there must have been food they could not reach. Also, safety in an isolated tree could be gained only after crossing open ground – often perilous in the presence of neighbours like the dinosaurs. The ability to search a wider area for food and to escape from predators were two advantages that would be gained by a new type of flying vertebrate.

However, there were already present some flying animals that had monopolized the air for nearly 200 million years. The air was full of juicy insects which were a potential source of food ready to be exploited by larger flying predators. Thus ecological selective pressures worked in favour of survival of flying animals for another time in the history of the Earth (these 'incentives' had already produced flying reptiles millions of years before).

Flight offered many other advantages:

- It provided the means of rapid colonization of new areas by enabling fliers to cross geographical barriers such as oceans and deserts.
- While it allowed escape from virtually all flightless predators, the birds could become supreme hunters because they could use the best vantage points to swoop quickly onto their prey.
- Because they could move long distances in a relatively short time, fliers could gather their daily needs from places that were far apart and could also follow the seasonally milder weather by migrating if necessary.

Warm blooded and feathery

In order to fly, animals must have the right shape.

- Flying animals need large surfaces which act as wings to provide lift and a means of propulsion.
- They must be streamlined or they will waste energy in overcoming drag. They must keep their weight to a minimum, for if they get too heavy they will not be able to become airborne in the first place.
- Powered flight of all but the smallest birds consumes energy rapidly and so flying animals need bodies that will meet this need by working at a high metabolic rate.

Birds have all of these adaptations. Flight has affected the shape and physiology of almost every part of a bird's body, but perhaps the greatest evolutionary stride was taken when scales, which had played such an important part in the success of the reptiles, evolved into **feathers**.

All birds keep their weight to an optimum according to their flight requirements. Their skeletons are light, partly because many of the individual bones are hollow but also because birds have shorter bodies and fewer bones than any other vertebrates. Modern birds have lost their reptilian tail; they have light skulls and beaks, and have lost their heavy teeth. Birds keep their weight down in other ways too. Their digestive processes are very rapid, so that they seldom carry excess weight in their stomachs and intestines. The reproductive organs weigh about one-tenth of the bird's total weight in the breeding season but at other times they may weigh one thousand times less. Eggs are laid soon after fertilization – this is an important saving because, in a small bird, a clutch of eggs may weigh more than the bird itself.

A bird's body must be able to resist distortion when the flight muscles contract, or energy would be wasted and the bird would deviate from its flight path. Birds overcome this problem by having very inflexible bodies, with many bones fused together. The best known examples are 'the wishbone', formed by the fusion of the collar bones in a way that keeps the shoulders apart when the flight muscles contract and the equivalent to our ankle bones, which are fused to give strength when landing and avoid dislocation.

Feathers are unique to birds. Like the body coverings of other vertebrates, they are composed of the hard-wearing protein called keratin. They are very strong and resilient, but exceptionally light. They also provide insulation and waterproofing, because they trap air really well. In this respect feathers far surpass the hair of mammals. As a result, small birds are much better at surviving lower temperatures than are mammals of similar size.

Flight demands a very expensive lifestyle in terms of energy requirements. The evolution of birds has only been possible because of some remarkable physiological adaptations which accompanied their anatomical modifications. Their biochemistry allows them to release energy from their food stores very rapidly and their feathers ensure the insulation that is essential if the energy is not to be lost as heat. Birds have special lungs with a system of air sacs which extend throughout their bodies. These make certain that there is always a high diffusion gradient between the air in the lungs and the blood, so exchange of gases is very efficient. The heart is comparatively large and beats at a high rate, thus pumping fuel and oxygen in the blood quickly to the muscles.

A bird's ability to fly is linked to its characteristic of being warm blooded, but warm bloodedness has contributed to other aspects of a bird's life. The young are not able to gather food for themselves, even though they require large amounts to maintain their high metabolic rate. Parental care is therefore obligatory. It begins with nest building and continues through incubation of the eggs to feeding the chicks until they can fend for themselves. This behaviour is in marked contrast to that of other egg layers like the reptiles and insects, which usually leave the eggs to hatch on their own or at best bury them for protection.

Early birds

There is no doubt that birds evolved from now extinct types of reptiles. Until the 1990s, palaeontologists were certain that the earliest bird was ***Archaeopteryx lithographica***. The name means 'ancient wing from the lithographic stone' and is derived from *Archaeo* = 'ancient', *pteryx* = 'wing', and *lithographica* = a type of fine-grained limestone used for a particular type of printing called lithography. When alive, it would have had a mass of about 270 g and a wing area of about 480 cm^2 – about the size of a magpie.

Archaeopteryx had a long lizard-like tail and beak-like jaws lined with teeth, but its collar bones were united to form a bird-like 'wishbone' and it had feathers almost identical to those of modern birds. It dates from the Upper Triassic, some 140 million years ago. A fossil feather was discovered in 1860 in the fine-grained limestone from Solnhofen, Bavaria, and it was given a scientific name a year later by **Hermon von Meyer**. In the same year, in the same area, an almost complete fossilized skeleton was found with feather impressions. This was bought from the quarryman who found it by a local doctor and amateur fossil hunter, **Carl Haberlein**. He soon recognized it as a potential money maker and advertised it to local scientists, cleverly not allowing them to make drawings or detailed notes of any kind. The fate of this fossil then depended on the highest bidder. As it happened, the curator of the museum in Munich, **Andreas Wagner**, was a Biblical fundamentalist who would not accept the idea of evolution and therefore did not recognize the value of the specimen. No money was forthcoming from that source! Haberlein eventually sold his entire collection of fossils to the British Museum in London, where it remains. Archaeopteryx is probably the most valuable and sought-after treasure among fossil

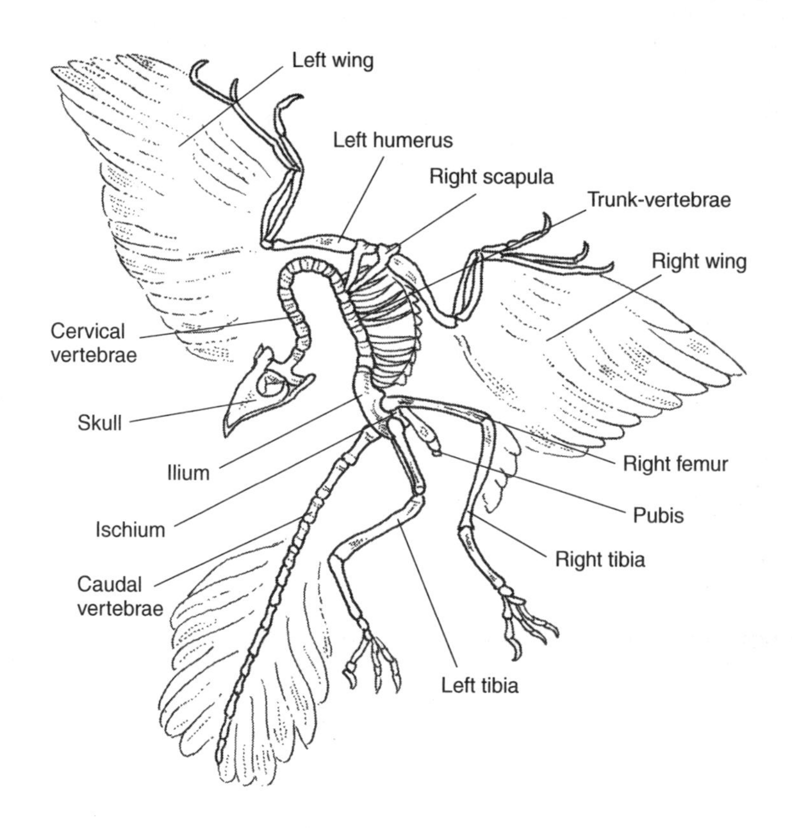

Figure 3.10 Archaeopteryx

collectors throughout the world; indeed, it has been called 'The Mona Lisa of palaeontology'. The price paid for Haberlein's collection was £700, almost a king's ransom in the late nineteenth century (in fact, it was twice the annual acquisition budget of the whole of the Natural History Department of the British Museum at the time). Richard Owen, of dinosaur fame, was the director and made the decision to purchase it. When you consider that Mary Anning was peddling her quite common fossil ichthyosaurs and plesiosaurs in 1820 for £150 or more, Owen had a bargain.

A second, and more complete, specimen was found in 1877 and purchased by Haberlain's son, Ernst. He put it on the market, but being more shrewd than his father demanded a much higher price than the original London specimen fetched. Teutonic national pride determined its fate this time. The Prussian State paid DM20 000 (£3000/$4500) for it and gave it pride of place in the Humbolt Museum, where it is known as the Berlin specimen.

It seems that during the Jurassic, muddy lagoons were common in the Solnhofen region of Bavaria and it was there that the ancient birds became trapped and eventually fossilized. The result of searching the region for 140 years since the first find has been the discovery of seven Archaeopteryx fossils. Compare this with a recent amazing occurrence in the Liaoning province in China where, in 1994, farmers uncovered hundreds of fossils of early birds. Chinese palaeontologists named one as the 'holy Confucius bird', ***Confuciusornis sanctus***, which could knock Archaeopteryx off its perch as the oldest known fossil ancestor of birds. These fossils also are about 140 million years old but, unlike Archaeopteryx, they possess a modern-looking beak and a bird-like tail to support tail feathers. In 1996 another fossil, **Protoarchaeopteryx**, was found in the same area, which threw doubt on Archaeopteryx being the only possible ancestor of true birds. By 1997 it was found that the fossils of Protoarchaeopteryx were up to 30 million years younger than those of Archaeopteryx and some palaeontologists suggest that it could be the ancient equivalent to modern flightless birds (like mini ostriches). Whatever is the outcome of debates about these new finds, all of these fossils have played a vital role in proving to early doubters of evolution that reptiles had indeed taken to the air.

The discovery of the fossil of Archaeopteryx came just two years after Darwin's publication of *The Origin of Species* and was hard evidence to confirm Darwin's idea that one group of animals developed into another by way of intermediate links. When Thomas Huxley championed Darwin's controversial theory, he also predicted that just such a creature must have existed and subsequently anticipated its discovery. He proposed that Archaeopteryx was a feathered dinosaur. If the impressions of feathers had not been so well preserved, Archaeopteryx would have probably been identified as a meat-eating dinosaur. There is still much controversy about details of the ancestry of Archaeopteryx but the most popular theory is that it evolved from a type of small dinosaur called a **coelurosaur**. These animals lived in the Triassic. They were bipedal, using their long legs to run quicker than their prey. A feature that they shared with birds was the arrangement of their toes. They had only four toes on each foot; three that pointed forward and one that pointed to the rear and which did not touch the ground.

Among the bird-like features that Archaeopteryx had not developed was a **keel** – a vertical downgrowth from the breastbone which supports the flight muscles in modern birds. Some ornithologists believe that without a keel, flight would have been almost impossible. In fact there are many theories of how flight evolved. One suggests that the ancestor of Archaeopteryx was a runner which eventually jumped into the air. The other is that it was a climber which became a glider before it began to fly. The fossils of coelurosaurs are so like Archaeopteryx that the two have often been confused. A fossil was identified as Archaeopteryx in 1971 after it had been in the museum in Haarlem, the Netherlands for more than 20 years labelled as a pterosaur. Perhaps it is the size of Archaeopteryx which offers a clue to the development of feathers and subsequent flight.

A bird which lives in South America, the **hoatzin**, has some primitive features that are very similar to Archaeopteryx. Apart from baby ostriches, it is unique among living birds in having chicks with movable claws on the ends of their wings, enabling them to hang onto branches. As the chicks grow the claws are lost and the wing feathers develop, but even the adults use the wings for climbing. So, with a little imagination, the life cycle of the hoatzin can be seen to repeat the evolution of birds, from the lizard-like creature climbing trees to a winged bird that can fly from one to another.

The coelurosaurs were the smallest of the dinosaurs, and Archaeopteryx was 30 cm smaller. Small animals benefit from being able to survive on small quantities of food. They suffer in there being a greater chance of them becoming the prey of larger predators and also from having a relatively large surface area. Animals with large exposed surface areas are much more likely to gain or lose heat as their environmental temperature alters. Perhaps this is why feathers evolved as a form of insulation.

Many reptiles have scales with a central stiffening rod or keel, and some scales have serrated edges, so scales had the potential to evolve into feathers, but what was the selective pressure necessary to produce this change? Although modern reptiles are cold blooded, it now seems likely that many dinosaurs were warm blooded – among these were the coelurosaurs. With their relatively large surface area it would be the small animals that needed insulation. Of today's mammals it is only the bigger ones, like the elephant and the rhinoceros, that can manage without a fur coat or a thick layer of blubber. Similarly, the giants among the dinosaurs could perhaps have survived without insulation but not the smaller ones.

Opponents to the theory that feathers were primarily important for insulation have pointed out that the climate was mainly mild during the Jurassic and early Cretaceous, but there *were* still times when it was cold. There were glaciers in some regions and it certainly became cooler at night or when winds followed rainstorms. In the tropics today small warm-blooded animals have coats which provide insulation. These are particularly important for animals that live in trees and which find it difficult to shelter from the frequent cooling winds.

A popular theory is that the birds evolved from reptiles that ran after insects. Feathers initially provided insulation but then some feathers on the arms were modified to form insect traps like nets. Eventually the ancestors of the birds began jumping to catch insects, then came flapping (to gain height), and eventually true flight. A problem with this suggestion is that, at a time when competition on land was intense, and when there were a large number of terrestrial predators, could an animal that relied upon running afford to get smaller and develop feathers? Both these developments would slow it down – and feathers would certainly decrease its chances of running between obstacles or hiding in holes or between rocks. Smaller animals have a greater need to find places to escape from predators. The only place left for an animal with conspicuous adornments like feathers would be up a tree. In trees such animals would also find greater safety for their eggs.

Once the ancestors of birds had mastered true flapping flight they must have spread quickly throughout the major continents. By the Cretaceous, birds had reached Australia, and in Spain a fossil feather has been found that is as old as Archaeopteryx. The next land bird that was found was **Alexornis**, a bird resembling a kingfisher. There is more evidence of water birds of that time which may be an indication of the extent of the Cretaceous seas – they covered large tracts of our modern land masses – or might reflect the greater likelihood of skeletal material being preserved in the chalk beds being laid down by the sea at that time. It might also indicate that very early in their evolution birds were attracted to the water's edge. This is not surprising because fresh water would have attracted lots of insects just as it does now. Here they would lay their eggs, young would develop and fly above the water. Other animals would feed on the adult insects but none would be as well equipped to catch them as the birds.

In the shallow coastal waters, sea shores provide rich feeding grounds for birds today and this has probably always been the case. Water birds had the advantage of being able to fly to new feeding grounds and eat animals on the edge of the water, or they could enter the water and swim to catch food.

It was not long before some birds became well adapted for swimming and catching fish. Fossils of birds which show adaptations for feeding on fish have been found in deposits dating back over 100 million years. A feature of these early fish-eaters was the presence of teeth. This was an adaptation for feeding, but for a long time the '**toothed birds**' were given a central position in the evolution of modern birds. Another group, which were of similar interest, were the **flightless birds** – the ancestors of modern ostriches, emus and kiwis.

At the start of the Tertiary (about 65 million years ago) there was a lack of terrestrial carnivores, and so birds could live on the ground in relative safety. A number of flightless birds evolved which looked as though they might become the dominant terrestrial group – until they were overrun and hunted to near-extinction by mammals. Flight is so expensive in energy that it seems to be lost as soon as it is unnecessary. This was particularly true of the extinct dodo and of the flightless cormorant of the Galapagos, which still exists and which Darwin described during his evolutionary studies.

Just as Nature abhors a vacuum, evolution abhors an empty niche. Thus, many large flightless birds occupied the ecological niche vacated by the dinosaurs and other terrestrial predators when they suddenly became extinct. Birds of gigantic proportions roamed the continents. Some of their bones have been found in Europe and North America, showing that they were often 2 m in height with massive beaks for flesh tearing. In 1917 the American palaeontologists **W.D. Matthew** and **W. Granger** described such an avian monster from fossil bones found in Eocene deposits in Wyoming. They called it ***Diatryma gigantea***, a fierce carnivore with a huge compressed beak and vestigial forelimbs. It must have bludgeoned and torn its prey into submission. New Zealand is the home of the first flightless bird fossil ever to be found. It was **Dinornis** – a monstrous 3.5 m in height.

The better known moas of New Zealand became extinct only in relatively recent times, hastened by their hunting by humans. Another well known extinct type is the 'elephant bird' of Madagascar. Remnants of its bones and egg shells are often found. These types are known only from the Pliocene (5–2 million years ago) and so they provide evidence of how rapidly flightless and massive increases in size can take place.

Another group of flightless birds evolved from the wing-propelled divers. Today they are represented by the penguins but in the past there were also the great auks, some almost 2 m long. Most of them became extinct as the seals and porpoises evolved during the Miocene period (24–5 million years ago). As with many of the flightless land birds, it was the spread of mammals by adaptive radiation that caused their demise.

4 ADAM, THE PECULIAR ANIMAL

Warm blooded and hairy

As vertebrates, evolution dictated that we would begin as fish and end as men or women, but in between we went through some profound revolutions. Amphibians managed to get out of water without really conquering land; the reptiles conquered the land but not the weather. The **mammals** conquered the weather and, counting their other achievements as well, needed no further basic revolutions to produce us as humans.

While reptilian monsters reigned, only a particularly careful observer would have detected other animals on the landscape. They were there, however, apparently constrained from expansion by the reptilian success, and probably living in danger, with every sense primed for the approach of a dinosaur. These were, of course, the other descendants of the coltylosaurs (stem-reptiles): modern reptiles, and the first mammals.

During the late Permian and early Triassic periods (about 245 million years ago), at about the same time as the ancestors of the dinosaurs were increasing their influence on the land, other types of reptile, the synapsid group (see page 94), was increasing its hold in many environments. They were to become the mammals.

An early innovation among the synapsids was the evolution of a form of temperature control. Some species developed a large web of skin along their backs like a sail supported by long extensions of the backbones (Figure 4.1). The high surface area of the sail, richly supplied with blood vessels, would have enabled these 'sailbacks' to absorb energy from the sun when they were cold and to lose heat when they were too hot. A relatively constant body temperature allowed the synapsids more activity, regardless of the temperature of their surroundings. Their metabolic rate became greater as a result of their more efficient locomotion, feeding, and breathing – a prerequisite for a mammalian mode of life. Because of this, these were called the **mammal-like reptiles**.

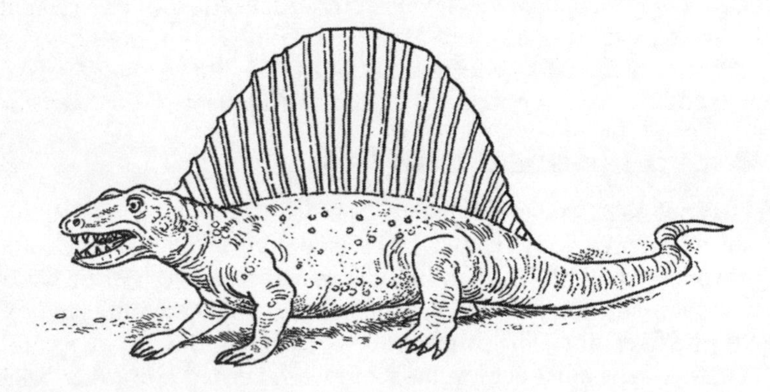

Figure 4.1 The sailback reptile, Dimetrodon

By the Middle Permian (265 million years ago), fossils show how limb bones had become long and slender, more efficient for locomotion. Forward movement on such limbs would require far less energy than the sideways sinuous movements of reptiles. However, fast movement requires a high fuel intake which was facilitated to some extent by adaptive changes in jaws and teeth.

Mammalian teeth vary in shape, each type being adapted for a particular function. Some are used for chewing, others for biting or flesh tearing. In contrast, reptilian teeth are the same shape and usually backwardly pointing so that they can gather and hold food before swallowing it whole. The process of digestion begins in a mammal's mouth by chewing and mixing food with saliva. In reptiles, digestion begins in the stomach and is therefore a slower process.

Faster food intake and its breakdown require better breathing to supply the oxygen to release energy for the digestive process. Judging by the position and structure of the ribs in some fossil mammal-like reptiles, it seems possible that they had a **diaphragm**, which is almost exclusive to mammals today. The development of a secondary **palate**, separating air and food intake, is another adaptation for regular breathing that is absent in primitive reptiles.

All these features characterize a high-energy system which is made more efficient by the maintenance of a constantly high body temperature. There is evidence in some fossil skulls of pits around the snout, which seem to be the same as those that house nerve endings in mammalian **whiskers**. **Hair** would almost certainly have evolved first, from reptilian scales, as an insulator and only later for specialized sensory mechanisms.

The mammal-like reptiles show an almost perfect series of transitional steps from the reptilian to the mammalian condition. Moreover, it is likely that many animals now considered on purely fossil skeletal evidence to be reptiles would be called mammals if they were alive today. Where, then, is the line between the two classes to be drawn? The arbitrary point at which the classes are separated concerns the structure of the **jaw**. The skeletal definition of a mammal is that it has only one bone present on each side of the lower jaw. If this strict definition is applied to fossils, creatures already present in the Middle Triassic period must be considered to be mammals. Also the middle ear bones of mammals are derived from jaw bones of their reptilian ancestors.

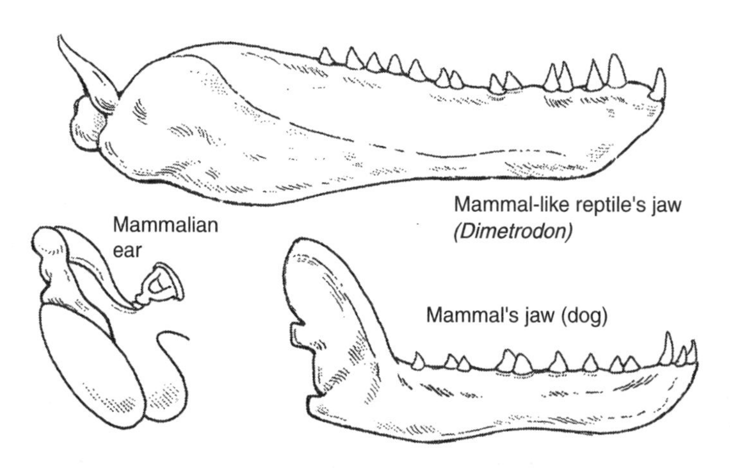

Figure 4.2 The bones of the middle ear in mammals have evolved from jaw bones of reptiles

By the middle of the Jurassic (about 190 million years ago) there were various groups of small mammals resembling shrews, with teeth suited to a diet of insects. They could probably hunt during the day and night because they were not dependent on the temperature of their surroundings to remain active.

All the earliest mammals were small – none larger than a modern domestic cat and most smaller than rats. The descriptions are based mainly on the arrangement and structure of their teeth. Early mammals diversified rapidly even during the long reign of the dinosaurs. Some became extinct relatively quickly but enough survived to inherit the Earth when dinosaurs finally disappeared.

From egg layers to live bearers

In spite of their small size, Mesozoic mammals (from 150 million years ago) were sufficiently variable for a number of groups to have been described. However, there was one small group that was so different from any of the others that scientists classified them as the early ancestors of present-day egg-laying mammals, the **monotremes**, which today include only the duck-billed platypus and the spiny anteater. Like the ancient mammal-like reptiles, these creatures possess a number of reptilian characteristics alongside undoubtedly mammalian ones. Besides egg-laying, the reptilian features include extra bones in the hips and shoulders and hardly any ability to control their body temperature. Despite these, they have the unique habit of suckling their young, although they produce milk from pores resembling sweat glands rather than mammary glands. They also have hair and a DNA profile which links them closer to mammals than to reptiles.

The other major groups are the **marsupials** and the **placentals**. They can be distinguished by minor skeletal features so that their fossils can be readily identified as being separate. However, the most important difference between modern types is the fact that the marsupial young are born at a very early stage of development; their growth, fuelled by milk, continuing in a pouch on their mother's abdomen. In contrast, the placentals retain the foetus until it is comparatively much larger, nourishing it during its growth via a special organ called the **placenta**. Both types are descended separately from an ancestral group. The fossils of placentals are known from deposits of equal age to those in which early marsupials appear. Today, marsupials have a discontinuous distribution throughout North and South America and Australasia. Their spread and survival in different parts of the world is very closely related to the break-up of ancient land masses and continental drift (see page151). After their origin, marsupials colonized large areas of the world but became extinct in Europe and North America, unable to survive competition with the placentals. South America separated sufficiently from the other land masses during the early Tertiary to exclude most placentals. The few that did arrive there before the separation failed to expel the established marsupials, one of the most remarkable of which was the South American **sabre-toothed marsupial**, which paralleled closely the sabre-toothed tigers of North America and Europe.

When the land connection was re-established between North and South America in the late Tertiary, there was a migration of animals in both directions. The sabre-toothed marsupials died out due to the onslaught of cat-like carnivores which competed with them in the same way that other placentals caused the extinction of most other types of marsupials. Only a few survivors, notably the ancestor of the Virginia opossum, made the move north.

From South America, the marsupials had also moved across what would later become Antarctica into the area which eventually became the island continent of Australia. There, the pouched mammals evolved into a wider range of forms, each adapted to its particular niche. Today's Australian mammalian fauna is a mixture of natural marsupials and alien placentals which were introduced by humans. The numbers and diversity of native marsupials have already been drastically reduced by the consequent uneven competition.

Towards the end of the twentieth century, this widely accepted theory about the evolution of Australia's unique mammals was overturned by the discovery of a tiny fossilized tooth no bigger than a pinhead. It was identified in the early 1990s as that of a placental, having been found in 55-million-year-old clay deposits in the back yard of a home 150 km north-west of Brisbane. Scientists had assumed that marsupials flourished in Australia because they never had to face competition from placentals, which they thought had been unable to spread to the continent from the north. The 55-million-year-old tooth from southern Queensland demonstrates that the placentals appeared in Australia at the same time as the marsupials, but that the marsupials prevailed. The tooth comes from an animal about the size of a rat and resembles teeth from mammals which lived in Europe and North America between 45 million and 70 million years ago. Until this find, the earliest known placental mammals in Australia were rodent-like animals that arrived about six million years ago. Using a scanning electron microscope, scientists have established that the enamel of the tooth is similar in structure to the enamel on the teeth of placental mammals but is unlike that from any known marsupial.

Placental mammals, apparently present from the time of the first dinosaurs, became the dominant land vertebrates after the end of the age of the reptiles. At first there was some competition from the newly evolved birds, but the versatility of their form, unhampered by specializations for fight, was decisive in the struggle for survival. By the early Palaeocene (65 million years ago), placental mammals were set on an evolutionary road towards diversity of shape and size which matched anything that had preceded them. Today, the age of mammals is past: they were on the decline even before the evolution of humans and, although many extinctions in historical time were due entirely to human manipulation of the environment, the picture as presented by the fossil record is of a richness of species which no amount of conservation could restore.

Before the end of the Cretaceous (65 million years ago) placental mammals had adapted to a variety of environments with the result that there are fossils of not only shrew-like types but also carnivores and primates. Their food was the abundant invertebrates – a source sufficient to power, in the early stages at least, the radiation of mammals that was to follow. The great herds of grass-eating herbivores which roamed the plains of the continents were still in the distant future.

The simple body plan of Cretaceous mammals has been retained in several living groups. In many cases they are exceedingly specialized for a particular way of life but most tend to have five fingers and toes and a collar bone, as well as a simple brain structure. The remains of insectivores differing little from those of today's animals have been found from the Tertiary, and most of the modern families of this group can be traced back to the Eocene period. There is no doubt that some of the mammals found food and shelter in the trees of the great forests which extended across much of the northern hemisphere at this time. Early **primates** resembling tree shrews and lemurs are known from the beginning of the Tertiary.

Although the mammals rapidly established themselves as the dominant land vertebrates during the early Tertiary (about 50 million years ago), they ruled only about 30% of the Earth's surface, the rest of which was taken up by the oceans or fresh water. It is not surprising, therefore, that mammals belonging to several different groups gradually became adapted to an aquatic mode of life. Marine niches, in particular, were available to them because the great marine reptiles had become extinct at about the same time as the dinosaurs. Fish were abundant in the late Cretaceous period, but most of these were small, leaving opportunities open to larger animals.

Water has many advantages as a dwelling place. It offers considerable depth as well as large areas in which an animal can move. Its temperature is more constant than that of air but because its density is about 800 times as great as that of air it restricts speed of movement. However, water can physically support much larger animals than can be supported on land – hence the largest animal that has ever lived is the blue whale. Furthermore, mammals' warm-blooded systems have allowed them to colonize extremely cold oceans, with the result that there are very few large bodies of water which do not support aquatic mammals of some sort.

By the late Miocene (20 million years ago) mammals which show distinctive adaptations to a diet of flesh of larger animals had evolved. Their fossils show biting and stabbing teeth, and some modified for scissor-like slicing.

Most of the carnivores living today still have these so called **carnassial** teeth: they have been lost only by those types that have changed to a largely herbivorous diet, for example bears. The earliest carnivores were small and may have been good tree climbers – suggested by their spreading toes and the fact that their first digit was opposed to the sole of the paw or foot. Two main groups of carnivores with toes diverged during the Eocene and Oligocene (37–24 million years ago). These were the cats, civets and hyenas on one hand and dogs, bears and weasels on the other.

At the opposite end of the feeding order, woody plants with hard but nutritious fruits and seeds provided a new source of food for those mammals with mouths and digestive systems that could cope. From the Triassic (245–202 million years ago) to the present day, many creatures have adapted to take advantage of this diet. Both gnawing types and grazing types have perfected feeding mechanisms to take advantage of the first link in food chains – the plants. Incisor teeth at the front of the jaws are **open rooted**, which allows them to grow throughout the life of the **gnawing mammals (rodents)**, because the abrasive action of the hard food wears heavily against them. The rodents are, numerically, the most successful group of mammals on Earth. About one-half of today's living mammals are rodents. Some species, such as rats and mice, rival humans in the size of their populations and provide aggressive competition for the same food resources. The earliest gnawing animals were mammal-like reptiles and were replaced by primitive rodent-like mammals in the Jurassic. Modern rodents supplanted these in the Eocene and many extinct species are known only from their robust teeth because the rest of their skeletons are so fragile that they tend not to survive as fossils.

Other specialists as plant-eaters are the grazing types of mammals. They tend to have hoofs and are called **ungulates**. They include cattle, sheep and horses. The larger bones of these types have fossilized well and so they are among the best known of prehistoric animals. There has been an almost parallel evolution of flowering plants and herbivorous mammals. Although no true ungulates have been uncovered from Mesozoic deposits, ancestral forms must have been present because hoofed animals appear in great diversity early in the Tertiary. The two major groups of living ungulates are the **odd-toed** and the **even-toed** hoofed animals. These two groups evolved along different paths in the tertiary. The odd-toed types included the ancestors of the horse, rhinoceros and tapir. Today, all these animals are on the brink of extinction in the wild, yet in the past they included a vast array of creatures. Indeed, the largest of all land mammals was a hornless rhinoceros which stood 6 m at the shoulder. The even-toed types were present in the Tertiary but expanded in the Miocene when the first **cud-chewing** mammals appeared. Camel-like types were probably present before then, but ancestors of cattle, antelopes, giraffes and deer evolved to take advantage of the abundant sources of plant food on the expanding grasslands. These animals have developed horny pads that can pulverize vegetation, to replace the upper front teeth, and developed stomachs with several compartments for efficient digestion. As a result of these adaptations, nearly all medium-sized and large plant-eating mammals are members of this dominant group, and have among the best chances of survival in the changing world.

A vast assemblage of fossil animals and plants has been recovered from Eocene and Lower Oligocene deposits in Fayum, Egypt, that give us clues about the structure and behaviour of the ancestors of many present-day mammals. Most important have been the earliest members of the **elephant** group. They were among the most successful and varied of large herbivores during the Tertiary. Today's elephants are distinguished from other mammals in having trunks which evolved from the flexible upper lip of their ancestors. They also have tusks, enlarged incisors that have undergone extensive modification. The ancestors of elephants spread throughout the continents except Australasia and some types, like the mammoth, developed heavy fur to protect them against the intense cold of the Pleistocene Ice age.

Meet our ancestors

The group of mammals to which you and I belong, together with our nearest animal relatives, is called the primates, the 'numeros unos', the 'number ones'. This seems only fair – we are peculiarly extraordinary animals and therefore our closest relatives should share the credit. An all-embracing definition of our group is difficult to make because, unlike other groups of mammals, primates have no obvious unique characteristics. Indeed, we retain many primitive features from our early ancestors – the small insectivorous mammals which lived alongside the dinosaurs 70 million years ago. Aliens from another galaxy, visiting our planet some 70 million years ago, would probably have collapsed in hysterical laughter at the thought of Palaeocene primates developing into humans. But they did!

The primates – from the most primitive to ourselves – show a gradual modification in certain features that can be linked to the ecological situations in which they live. In the rat-like tree shrews, for instance, the eyes are located at the sides of the head, providing good all-round vision for detecting their insect prey and avoiding predators. But in the lemurs, monkeys and higher primates the eyes are at the front, giving stereoscopic vision and allowing good judgement of distance – important in animals that move rapidly through branches, and an inherited feature of humans. Good vision is more important than a sense of smell, and very keen eyesight has evolved in many nocturnal species of primates.

Certain other evolutionary trends are common to the group as a whole, often also related to their life in trees. For example, the skeleton is relatively primitive and, unlike that of the early mammals of the Cretaceous, is made up of a large number of separate bones. It includes five 'fingers' and five 'toes' on each hand and foot; the thumb and, except in humans, the big toe can be opposed to the other digits to grasp and make precise movements for feeding and grooming. In many primates the claws of primitive mammals have been replaced by flat nails supporting sensitive finger or toe pads.

Because the diet of many primates is not specialized, their teeth are also rather unspecialized and are large in number. The early mammals of the Cretaceous period had 44 teeth, and most lower primates still have 36, whereas humans and our closest relatives have 32 teeth. These are large numbers compared with other animals, such as rabbits with 26 and rats with 16.

Primates have also evolved complex **social systems**, enabling members of the group to keep in touch with each other in dense forest by the use of signals such as special calls and scent-marking. Even the most primitive primates have relatively small litters of young and spend more time caring for their offspring than most other mammals. The social life of primates has been one important stimulus to the development of a relatively large and complex brain. Humans have continued, and in some ways developed to a higher degree, some of these trends, including the sensitive grasp of the hand, stereoscopic vision, loss of a keen sense of smell, complex social life with developed communication systems, good infant care, and above all the most complex brain in the solar system (and possibly in our galaxy).

Some of the earliest primates lived in North America and Europe about 60 million years ago. They were squirrel-like in appearance with a long snout and a good sense of smell, while the brain was very small and simple. By the Eocene (about 50 million years ago) creatures resembling modern lemurs had developed. However, none of these primitive types are likely to be the ancestors of higher primates, the **monkeys** and **apes**. Today the higher primates are represented by many types of monkey, known collectively as **New-World monkeys** and **Old-World monkeys**, and by a few types of ape. The New-World monkeys are those found in Central and South America, and the Old-World types are found throughout Africa and Asia. The apes are the '**anthropoids**' and include gibbons, chimpanzees, orang-utans, and gorillas. Genetic profiling shows that chimpanzees are our closest living primate relatives.

By the Oligocene period (37–24 million years ago) the New-World monkeys and the Old-World monkeys had already begun their separate evolutionary paths. Apart from fossils found in Fayum, Egypt, there is not much evidence of the origin of the Old-World monkeys and apes at this time. The Egyptian evidence shows a number of small 'proto-apes' living in a subtropical forest environment.

During the succeeding Miocene period (24–5 million years ago), apes were far more common than monkeys, a situation which is completely reversed today. It seems that in the early Miocene in Africa, although the apes were all tree-dwelling, fruit-eating forms, their range of sizes was much wider. While a species of **Proconsul** resembled small gorillas or chimpanzees in stature, others were smaller. Overall their skeletons were monkey-like, and they mainly climbed in the trees on all fours rather than swinging below branches, or 'knuckle-walked' on the ground, like modern African apes. The story of Proconsul's discovery began in 1927, when **H.L. Gordon**, a settler in western Kenya, found some fossils in a limestone quarry. He thought they might be important and sent them to Dr **A. Tindell Hopwood** of the British Museum. In a fist-sized nodule of limy matrix lay an upper jaw bone of a human-like primate. After raising money for an expedition, Hopwood went to Kenya in 1931, where he collected additional **hominoid** fossils and stated his conviction that the jawbone that he was sent originally was from a new genus ancestral to the chimpanzee. The creature's 'christening' came about 18 million years after his death when he was named Proconsul by Hopwood – with some sense of whimsical humour, he named him after Consul, a chimpanzee in a London vaudeville act that entertained audiences by riding a bicycle on stage while wearing a suit and smoking a pipe.

The next chapter of the story was written by **Louis** and **Mary Leakey**, who made a series of expeditions to western Kenya in the 1940s and early 1950s. They found the most famous Proconsul remains of all – a fairly complete skull. Since then Proconsul's remains have been recovered in abundance from the lower Miocene stream beds in East Africa. Many of his fragments probably came to rest when a crocodile had him for breakfast and spat out the few parts it could not chew. In making their discovery originally, the Leakeys had unearthed the skull of the first Tertiary ape ever to be found. However, the most complete specimens of all were preserved after the misfortune of being washed into a rapidly silting gully and drowning in the ancient Miocene watery grave. At least nine complete skeletons have been found, representing an age range from very small babies to adults. Analyses of Proconsul's abundant remains have indicated that it was *not* an ancestor of the modern chimpanzee or gorilla as stated by Hopwood. Instead, it appears to be a generalized ancestor of all the larger anthropoid apes and humans.

The way we were

About 17 million years ago, the African continent became connected with Asia, leading to the migration of early apes from Africa into new environments. Some three million years later, new types of apes had evolved in Africa, Asia and even in Europe. From the point of view of human evolution, the African types are the most important because their cheek teeth show particular similarities to our own, being relatively large and covered with thick enamel. This characteristic separates humans from the gorilla and chimpanzee. Of particular note is ***Ramapithecus***, which at one time was thought to be the ancestor leading to humans. However, it is now thought that another type, ***Kenyapithecus***, may be the common ancestor of the gorilla, chimpanzee and ourselves.

There were some evolutionary experiments that seemed to lead to dead ends. One of these was an enormous ape, ***Gigantopithecus***, which lived in China as recently as a half a million years ago. It was 3 m tall and was first identified from its fossilized teeth. In 1935, the eminent Dutch Professor **Gustav Heinrich Ralph von Koenigswald** bought a job lot of fossil teeth in a Hong Kong apothecary shop where such 'dragon's teeth' are sold to be ground up for medicinal purposes. Four unusually massive human-like teeth were among the collection. The teeth puzzled anthropologists for 20 years, becoming the basis of a mythical monster reputed to be 12 ft (3.6 m) high. Incidentally, it is one explanation for the legendary 'Abominable Snowman', the Yeti! In 1955, Chinese scientists decided to solve the mystery of this beast and sought the co-operation of merchants of dragons' teeth, who had crates of them in warehouses ready for a lucrative market. They found no less than 47 teeth and two whole lower jaws of the now famous *Gigantopithecus*. Most of these found their way to the Netherlands due to some Dutch cunning. Von Koenigswald suffered some complicated and unpleasant experiences during the Second World War – German born, but a Dutch national, he was separated from his family and put into a concentration camp by the occupying Japanese. It was mistakenly rumoured that he met his death there though his fossils enjoyed a better fate, despite the fact that the Japanese military considered them to be the property of their expanding Empire of the Sun and ordered them to be impounded. However, von Koenigswald got the better of his captors by persuading his Swedish and Swiss friends to put the teeth in large milk bottles and bury them. What the Japanese got were fakes, beautifully made out of brickdust and plaster so that even an accidentally broken piece would still look like a proper fossil. After the war von Koenigswald was reunited with his family and his teeth, which were returned to his native land.

About 6 million years ago, the ancestral line leading to the African apes (gorillas and chimpanzees) and humans divided. The evolutionary separation almost certainly took place in Africa, probably in forests or woodland environments. Early discoveries in Java and China pointed to the possibility that hominids first evolved in Asia. However, it has since been shown that this is not the case and that Africa was the cradle of human evolution with subsequent migration to the rest of the world.

In 1891 **Eugene Dubois** found fossils in Java of an early hominid, ***Homo erectus*** – **Java Man** (literally, 'upright man') – now known to be 1.8 million years old. The discovery was a fascinating example of an almost arrogant confidence. In 1877, aged only nineteen, Dubois began to specialize in anatomy and natural history at the University of Amsterdam. A few years later he became obsessed with the idea of finding a really primitive ape-like fossil human and concluded that the East Indies might be the place to find it. His decision to search there was probably influenced by a role model of his, **Ernst Haeckel**, the prominent German biologist of the day and an admirer of Charles Darwin. Haeckel had proposed, in theory, a new species of fossil that was half human, half ape, and without speech. He bravely gave this unknown beast the name ***Pithecanthropus alalus*** (*pithec* = 'ape'; *anthropus* = 'man'; *alalus* = 'without speech'). Haeckel also, mistakenly, believed that Asian apes were more closely related to humans than were the African apes so he confidently predicted the missing link to be in Asia. Firmly believing in Haeckel's hypothesis, Dubois approached the Netherlands government to sponsor an expedition to Dutch territories in the Far East. Not surprisingly, the prospect of financing a professor's assistant to find a purely imaginary creature was not very high in the list of the government's priorities. Their answer was brief – 'No'. However, the undaunted Dubois signed up as a medical officer in the Dutch Army. At the age of 29, after resigning from his comfortable job as a lecturer in anatomy in the University of Amsterdam, he travelled to the Dutch East Indies with his wife and child. There he was to make one of the greatest discoveries in the history of humanity. He was to find his 'missing link'.

In any 'Garden of Eden', searching for the remains of Adam or Eve is very much like looking for the proverbial needle in a haystack – or perhaps a million haystacks! In most cases their discovery would be accidental. Human bones are too fragile to be preserved as easily as those of many other larger animals and then again, even the most primitive humans were more intelligent than the animals about them and generally had enough sense to avoid getting trapped in bogs and quicksands where their bones could be fossilized. Of course, occasionally there were accidents and the more accident-prone primitive humans possibly fell into rivers and became silted over by river drift deposits. However, the chances of this are remote; the chances of fossilization even remoter; and the chances of finding such a fossil puts us back among the needle in the million haystacks game. If you put all the fossils of people-apes ever collected into one pile, it would hardly fill two average-sized rooms.

About 1.8 million years ago a Java people-ape met the fate that eventually befalls us all, except that his bones were destined for greater things. Totally unsuspecting, he became the most famous and most discussed member of the early human family. He was to fill a vacant niche in the Hall of Fame of Human evolution and now provides the ubiquitous bust in museums throughout the world with the carved letters 'The Earliest Known Man'. When he died his mortal remains were obviously not eaten by the abundant scavengers of his day. It is possible that they were smothered in a mud flow from a flash flood and came to rest on the banks of the Solo river in Java (now part of Indonesia). His flesh rotted away and his skeleton became dismembered but enough of his skull remained protected under a duvet of sediment to become fossilized. There it lay until August 1891, when Dubois wrote a whole new chapter in the history of human evolution by finding him – on purpose! In fact, labourers from Dubois' crew made the actual find and he was not present at the time to oversee his first fossil brought to light. Dubois suggested that the remains of the skull were too large to be that of an ape and too small to be that of a human. In May 1892 a femur was found only 13 m (45 ft) away from the first find and Dubois firmly believed it was from the same individual. His diagnosis of an upright human walker with a primitive cranium is now reason to call the creature *Homo erectus*, although it was first named *Pithecanthropus erectus* – literally ‘upright ape-man’. Incidentally, the suggestion to change *Pithecanthropus* to *Homo* was made in the 1940s but the sciences of palaeontology and taxonomy move by evolution rather than by revolution and so the official change was not made until 1960. The whole episode was a brilliant, astonishing triumph. It is almost the same as thinking that you are certain to win first prize on a National lottery and succeeding!

Dubois returned to Europe in 1895 and announced his findings at a meeting in Leiden, the Netherlands. Here was the first really primitive kind of human, a creature so long talked about in the abstract but now seen, if not in the flesh, in the bone. Then the sceptics began their merciless attacks and openly mocked Dubois’ opinion of his fossils. During the next 40 years poor old Java Man underwent some embarrassing moments when his parentage was questioned in public. No less than nineteen authorities on people-apes from a variety of European countries gave their varied opinions on his nature. Very few believed that the fossil fragments were from the same skeleton. By the end of the nineteenth century at least 70 books and articles had been published discussing Dubois’ Java Man and hardly any agreed with his interpretation. To the anti-evolutionists, of course, the whole thing was an illusion. They still believed that Adam was our first ancestor and that he had lived only 4004 years before Christ. Dr John Lightfoot, Chancellor of Cambridge University in England, even announced the day and hour of his birth as ‘23rd October, at nine o’clock in the morning’! Some said that the skull was that of a deformed idiot! On the other hand, certain proponents of the idea that it was a ‘missing link’ went to the other extreme – describing what his face was like, the colour of his skin, how much body hair he had and even what he ate for breakfast!

As a result of these attacks, the battle-weary Dubois began to feel dejected and took his fossils ‘off the market’, withdrawing them from scientific contact and becoming somewhat of a recluse. He took them home in a box which he kept under the floorboards of his dining room and ate his meals above the Java Man for many long years because he was so worried that jealous adversaries might steal them. Eventually he was persuaded to put the remains in the Teyler

Museum, Haarlem, his home town. There he locked them in a small safe within a larger safe. It was not until 1923 that the valuable contents were allowed to be examined by some of the leading anthropologists of the day.

All the information relating to Java Man stemmed from Dubois' specimens until 1931. Then the Geological Survey of the Netherlands Indies began excavations near the Solo river and by the turn of the decade had unearthed many examples of fossilized people-apes that could be related to Java Man. All the scientific fraternity was thrilled by the new discoveries – but, strangely, with the notable exception of Dr Dubois. He tried his best to discredit the new finds but failed. Perhaps the weird ideas that made him want to incarcerate his specimens for more than a quarter of a century influenced him again. Dubois died in 1940 and on his grave is carved a skull and crossbones, representing his life's work. He will be forever honoured for his prophecy and energy in finding the *original* Java Man.

Our man in Peking

Dubois' Java Man was more fortunate than his Chinese cousin, **Peking** (now Beijing) **Man**. At least the former was kept safely under Dubois' floor boards; Peking Man seems to have disappeared as one of the first casualties of the war in the Far East in the 1940s, over a million years after he had died in the first place. You might think that Dubois' experiences would have put off others with the ambition to find the missing link, but this was not the case.

In 1919 **Davidson Black**, a young Canadian physician, took a job at the Peking University Medical College teaching neurology and anatomy. His experiences in China were strange indeed. His teaching demanded human bodies for dissection and so he applied to the local police for a supply. They obliged by sending him executed criminals but minus their heads. After complaining, he was sent living criminals and told to execute them himself! Needless to say there was a limit to what he would do in the name of education and he sought other arrangements.

The story of Peking Man goes back to 1903 when an ancient human tooth turned up in a Peking apothecary store and was bought, but then lost, by the celebrated German palaeontologist, Professor **Gustav Schlosser**. Schlosser predicted that in the future someone would find the remains of some human-like ape that had lived in China during the early part of the Ice Age. By 1921 there was a gathering of fossil hunters working near the village of Choukoutien (now Zhoukoudian) during the Geological Survey of China. Prominent palaeontologists of no less than seven different nationalities pooled their various skills in the quest for the distinguished dead. Until the Swedish geologist, Dr **J. Gunnar Andersson**, came to Peking in 1914 as mining adviser to the Chinese government, little attention had been paid to the fossil wealth of China. Fossil bones and teeth were considered to be efficacious medicinally and used together with tigers' testes, bats' brains and brigands' blood to cure everything from dandruff to diarrhoea. Even today very strange things are used in Chinese medicine, and will probably continue to be used for ever more. In this way hundreds of thousands of scientifically priceless specimens must have been ground down to powder and sold with a promise to cure a vast variety of ailments over the years. Black and Andersson, with Swedish finances, were determined to make a systematic study of the treasure trove of fossils that was being unearthed in local mines and other excavations. They encountered quite a few problems along the way; not the least important being the Chinese belief in *feng-shui*, the spirits of the earth, wind and water which guard all burial grounds. Some places in China are so thickly populated that it is very difficult to find an area free from cemeteries. The fossil hunter must not offend local custom by digging on such sacred ground. Once, after Andersson had cut through the red tape that encumbered his plans, his operations were halted despite his granted permission to dig. An irate elderly woman was so enraged that she squatted in the hole that the palaeontologists had dug and refused to leave. The only way that he could persuade her to go was to hold an umbrella over her and offer to take her photograph. At the sight of the camera she leaped out of the hole screaming aloud. She continued to create such a disturbance that Andersson agreed that discretion was the best part of valour and left her to it.

Black and Andersson moved their excavations to 'Chicken Bone Hill', 40 miles south-west of Peking. Bits of quartz which did not belong to the geological stratum in which they laid were found in a large cave. Andersson became curious about the discovery. Why was this find considered to be so unusual? He reasoned that because animals do not use quartz, or eat it or wear it, this particular mineral had no business being in a limestone cave and had no natural way of getting there. Fossilized people are never obliging enough to have their tools in their hands, but quartz is ideal for making stone tools and this gave a clue to the former presence of prehistoric inhabitants of the cave. Andersson is quoted as saying 'In this spot lies primitive man. All we have to do is to find him!' After further wearisome digging Black found two human teeth but the world as a whole did not get particularly aroused by these despite the fact that he took them on a world lecture tour. In fact, the media, which usually feed ravenously on anything do with 'missing links', gave it minimal coverage. A couple of teeth are hardly material for front page news.

By 1929, more remains of the cave's inhabitants were found fragment by fragment. In the course of 10 years, from 1927 to 1937, Black, Andersson and the Chinese Professor **Pei Wen-zhong**, recovered bits of forty individuals of both sexes, among them the skeletons of fifteen children. All the bones showed ghastly evidence of cannibalism. The cave must have been a smelly place indeed in those prehistoric days. The fact that only seven bones were from any part of their bodies other than the head suggested that Peking Man did his killing and dismembering outside the cave and took only the heads of his unfortunate victims inside the cave for home consumption. There he cracked open the skulls to partake of the brains. A good many pieces of face and jaw were found, but not attached to the brain case. Thus, the scraps of the very first Chinese take-away meals have provided anthropologists with a feast of evidence of the behaviour of our man in Peking. Apparently, they had tools to chop up their colleagues and fire with which to cook them. Incidentally, of the 148 teeth and 13 lower jaws that have been found and examined, there has been no evidence of any dental caries, pyorrhoea, or other dental troubles – all this without the benefit of toothpaste! More and more Peking people picked and probed at the brains and marrow of their colleagues before littering their home with bone fragments for many thousands of years. The floor of the cave rose higher and higher until earth and bones filled it completely. A gruesome mixture of filth, bones and soil expanded to a height of over 30 m (100 ft) and a length of 150 m (500 ft). Despite their aversion to cannibalism, today's anthropologists are grateful that our ancestors were unwittingly packing away their bits and pieces for future fossilization and for science.

In 1929 Black announced Pei's find of part of a fossil skull to the Geological Survey of China. He named it *Sinanthropus pekinensis*, 'Chinese man of Peking', but together with Java Man he is now called *Homo erectus*. Black's destiny was a happier one than that of Dubois. He was hailed as the finder of the missing link and was made a Fellow of the Royal Society in 1932. Some say that he worked himself to death. His lifestyle was to work through the night until the early hours and then sleep until midday. This, and the probability that he also suffered from silicosis from so much rock drilling, proved to be his downfall. He was found dead at his desk in 1934, not 50 years old. Dubois found the missing link but gave up his scientific credibility; Black found the missing link but gave up his life.

Black's successor was **Franz Weidenreich**, a German anatomist who was forced to leave his professorship in Frankfurt in 1934 because he was Jewish. At 61 years old he became an American citizen and replaced Black in the University of Peking. Between 1936 and 1937 he made many finds of jaws and skulls of Peking Man but unfortunately history of another sort began to interfere with his progress in anthropology – the Japanese took over North China in 1937 and showed mounting interest in local fossil people-apes, which they evidently thought had become Japanese citizens due to the extension of their Empire into China. However, the director of the Chinese Geological Survey, Dr **Weng-Wenhao**, had other ideas.

In 1941 it appeared that the Japanese were about to carry off the whole fossil tribe to Tokyo so Weng-Wenhao decided to send the remains to America. In the meantime Weidenreich had made plaster casts of all the fossils of Peking Man and had taken them to the USA. The originals were packed in crates and loaded as 'top secret' cargo on a special train under the guard of nine American marines.

The train was meant to meet the liner *President Harrison* in the port of Chinwangtao (now Qinhuangdao) and it seemed that Peking Man was due for VIP treatment with a safe passage to the USA. However, unfortunately for him and for the science of anthropology, the train arrived in Chinwangtao on 7 December 1941. In the words of President Franklin Roosevelt, that was 'a day which will live in infamy' because it was the date that the Japanese bombed Pearl Harbour and brought the Americans into the Second World War. The fate of our man in Peking is a mystery from this point on. Historians have recorded what happened to the *President Harrison*: it was grounded to prevent its capture, but then refloated and used by the Japanese under the name *Kachidoki Maru*, only to be sunk by the American submarine *Pampanito* in 1944. The nine marines were captured when the train was intercepted by Japanese soldiers and imprisoned for the rest of the war, but what happened to their charge is anyone's guess. Rumours abound and include stories of the bones being ground up for Chinese medicine, and that they were actually put on a ship which capsized in the harbour. Certainly, they did not reach the USA. It is quite likely that Peking Man never left the shores of China. When the train was captured it was looted and poor old Peking Man was probably considered just a pile of old bones by the Japanese soldiers involved. They might have been the real raiders of the lost link – after all, could you really expect them to be anthropologists like Indiana Jones in *Raiders of the Lost Ark*? Perhaps the precious fossils ended up in a rubbish bin or were dumped into the dock. In any case, on the scale of atrocities which took place between 1941 and 1945 in the Far East, the whole episode was arguably of minor importance. Since the end of the Second World War many have searched for the lost link but to no avail. Imagine how Weidenreich felt – his claim to infamy was that he was the man who lost the missing link. Despite this blemish in his record, he made a positive contribution to anthropology by accumulating a synthesis of fossil humans and putting them in an evolutionary sequence. He died in 1948, adamantly believing that Java Man and Peking Man were races of the same species. In fact it took 15 years for the scientific world to call them both *Homo erectus*.

To the pioneering hunters of hominid fossils, it seemed that the 'missing link' in the form of Java Man and Peking Man had originated in Asia but it is now known that they came out of Africa. In 1961 Louis Leakey had found part of a cranium of *Homo erectus* in Africa but did not recognize it as such. Between 1984 and 1988, his son, **Richard Leakey**, and **Alan Walker** were the leaders of a team that found more remains of *Homo erectus* than anyone had ever seen before. Previous discoveries of this species had amounted to the odd scrap of arm or leg bone together with a jaw or two and some bits of skulls. Leakey and Walker found an almost complete skeleton at Nariokotome, on the west side of Lake Turkana in Kenya. What is particularly remarkable is that the fossil bones were all from one individual; an adolescent boy who lived about 1.5 million years ago. They christened him 'Nariokotome Boy' but he is sometimes called 'Turkana Boy'. The experience of Weidenreich served as a warning to Leakey. He was determined that his African missing link would not follow Peking Man to be lost for ever and so insisted that the Nariokotome Boy would never take any journeys away from the security of the National Museum of Kenya, Nairobi in which he is housed. Experts must visit *him*.

Out of Africa

Some of the most important fossils of our ancestors to be seen anywhere in the world have been found in Africa but right up to the 1990s religious fundamentalism prevented South Africa from becoming the focal point in the story of human evolution. In South Africa, evolutionary theory is still extremely controversial. The fundamentalist views of the Dutch Reformed Church have reigned for more than 300 years, and led to a ban on Darwinism being taught in state schools up to the late 1990s. This taboo on evolution is even more unfortunate because South Africa is one of the world's richest sources of hominid fossils – indeed, it was the site of one of the most significant discoveries of fossil evidence of human evolution this century. In 1924 the Australian-born anatomist **Raymond Dart** found the fossilized skull of a child in a box full of pieces of limestone. The fossils had been collected by the mine manager of a quarry at Taung ('the place of the lion'), on the north-eastern edge of the Kalahari. Dart was just about to leave his home to act as best man at a friend's wedding when he took a cursory glance at the box of limestone fragments and fossils that had been delivered to him during the day. He was amazed to see a fossil which he later claimed to be the 'missing link' between apes and humans and, therefore, the direct ancestor to ourselves. It was to become known as 'Dart's Child'. At the time, he was written off as a heretic because the prevailing view was that the first step in the evolutionary divergence of humans from apes was the development of a large brain. This theory was underpinned by the curiosity known as 'Piltdown Man' found near a road at Piltdown Common, Fletching (Sussex) in 1912. Supposedly a skull of one of our earliest ancestors, it was 40 years before this was proved a hoax that had been fabricated from a modern human cranium and an orang-utan's lower jaw. The features of the Taung child were the reverse: it had a small cranium and human-like jaws and teeth. Before Dart's discovery Asia, not Africa, was considered to be the cradle of Mankind but Dart and Louis Leakey destroyed this myth. The anthropological establishment of the 1920s dismissed Dart's claim that the Taung child, ***Australopithecus africanus*** ('African ape of the south') had anything to do with the origin of humans. Dejected and rejected, Dart virtually gave up research on hominids. As time went on, the limestone at Taung eventually ran out and no further trace of the fossil child or its parents ever came to light. Interestingly, in the 1990s it was suggested that the child had been carried there by a prehistoric bird of prey when markings on the skull were identified as possibly being made by grasping talons or a piercing beak.

The 1920s rolled into the 1930s and then there was a resurgence of interest by Dr **Robert Broom** (1866–1951), a Scotsman who was a country doctor working in the remote outposts of South Africa. He had helped Dart identify the Taung child's skull and was a palaeontologist of repute, having gained a medal from the Royal Society as recognition of his work on fossil reptiles in South Africa. Broom was encouraged to search for further evidence of *Australopithecus* after becoming curator in Pretoria's Transvaal Museum. Where should he begin?

There were plenty of limestone caves at Sterkfontein, near Johannesburg, so he drove out to have a look at them one Sunday in 1936. Broom learned that the quarry manager, Mr Barlow, had once worked at Taung, and that he was saving fossils to sell to tourists.

The owner of the quarry wrote in a little guide book to the places of interest near Johannesburg 'Come to Sterkfontein and find the Missing Link' – a strange prophecy indeed because Dr Broom did just that! Armed with some hard cash and an eye for a bargain, Broom went to work on Barlow, promising him rich rewards if he came up with the right goods. On 17 August 1936 Barlow handed him a fine brain cast and asked, 'Is this what you're after?' Broom's answer was a polite but not too enthusiastic 'Yes'. During that day and the next Broom found the impression of the top of the skull as well as its base, with parts of the forehead and side walls. One can detect more than just a slight note of satisfaction in this quote from him at the time of this discovery:

> To have started to look for an adult skull of Australopithecus, and to have found an adult of at least an allied form in about three months was a record of which we felt there was no reason to be ashamed. And to have gone to Sterkfontein and found what we wanted within nine days was even better.

Blasting of the limestone continued and Broom kept up his searches for the next two years but nothing appeared that would match the first skull to be found there. The nearest thing of interest was the knee end of a thigh bone.

Then in 1938, Barlow handed him an upper jaw with one tooth in place. To Broom it was another type of man-ape but he was curious because Barlow would not tell him where he had found it. It turned out that the fossil was not from Sterkfontein but was given to Barlow by a schoolboy. After giving Barlow a good talking to, Broom set out to find the boy, who he discovered at school 'with four of what are perhaps the most valuable teeth in the world in his trouser pocket.' The boy, Gert Terblanche, took Broom to a hill at Kromdraai, less than a mile from Sterkfontein, where he produced a piece of lower jaw which he had whacked out of its limy surround with a hammer and hidden there. The Kromdraai skull was named by Broom as *Paranthropus* but it was eventually decided that it was the same kind of people-ape as the Taung and Sterkfontein specimens.

The finds showed more clearly the features that distinguish humans from apes; for example, the molar teeth not present in children and the femur which provided clear evidence that *Australopithecus* had walked upright. The discoveries vindicated Dart's original hypothesis.

Two years later Barlow died, Sterkfontein was shut down and the Second World War began. After the war the persistent Dr Broom renewed his search. Broom's major claim to fame took place in 1947 with his discovery of a fossilized apewoman. He gave her the generic name *Plesianthropus* and described her as a million-year-old elderly female. Nicknamed 'Mrs Ples' by the media, she caused somewhat of a 'centrefold' interest in the local press but was dismissed by the scientific fraternity of the day along with the Taung child as 'just another ape'. The amazing Dr Broom died in 1951, aged 85, and still an eager and enthusiastic fossil hunter to the end.

There was an increase in interest in the claims of Dart and Broom when the mid-1950s exposed the infamous Piltdown Man hoax. Doubts were still expressed regarding the significance of the South African discoveries, allowing more attention to be given to Louis Leakey's and his wife's work in the Rift Valley. A significant breakthrough came in 1976, when Sterkfontein became the site of the discovery of the first specimens of the genus *Homo* to be found outside East Africa. Louis Leakey had discovered ***Homo habilis***, ('handy man') at Olduvai Gorge in Tanzania in 1959. He recovered an almost complete skull in excellent condition. The Sterkfontein fossils were undoubtedly *H. habilis* and provided evidence that South Africa's fossil people – apes of the genus *Australopithecus* – were part of the main trunk of human evolution.

Sterkfontein has yielded the secrets of more than 500 fossilized hominid specimens, making it one of the richest sites for fossil evidence of human evolution in the world. In 1991 archaeologists uncovered collections of stone tools and linked them firmly to *H. habilis*. It was the first South African evidence of the Oldowan stone tool culture, named after the Olduvai Gorge, where tools were first found in the 1930s.

In the beginning, the ancestral apes and ancestral humans would have looked very similar to each other. In deposits four million years old, however, fossils of a creature which is more closely related to us than to any ape are present. These first hominids were ape-like in most of their physical characteristics but were distinct in one very special way – they were bipeds; that is, they walked upright on two legs as we do. Something had been favouring the erect posture for millions of years before these men and women were born despite the fact that we pay for it, to this day, in terms of difficult childbirth, hernias, potbellies, and a hundred-and-one back problems. Why the change to walking upright occurred is conjectural; wading through water, looking over long grass, carrying food back to a home base, the need to cover long distances between trees using as little energy as possible have all been suggested as explanations. Whatever the reason, once this bipedal way of walking developed, various advantages were certainly available to our ancestors. In particular the hands were freed for use in various tasks, eventually to include the use and manufacture of tools. However, there is no evidence of the use of tools or of the existence of home bases from 4 million years ago.

Lucy in the Sky with Diamonds

Most palaeontologists classify hominids within one group called Southern apes (*Australopithecus*). Some of their earliest remains have been found in Ethiopia and Tanzania, dating from 3–4 million years ago. This creature bore many resemblances to modern chimpanzees but human features include the spinal cord extending from below the brain rather than pointing backwards like those of apes. The teeth showed a mixture of ape-like and human-like forms. One of the most famous specimens of this species is the skeleton known as **'Lucy'** found in Hadar, Ethiopia in 1974. She is a collection of bits of fossil bones made by **Donald Johanson** and **Tom Gray** and from the waist down appears to be human-like with legs and hip girdle suggesting the possibility of bipedal walking. Above the waist, she appears more ape-like, with long arms and a sturdy rib cage, hinting at tree-climbing habits. After gathering together her various scattered parts, Johanson and Gray were so elated that they sat up all night drinking beer and playing a record of an old Beatles favourite *Lucy in the Sky with Diamonds* – hence her more boring formal anthropological label, AL 288-1, was changed and she became 'Lucy'. She was classified as ***Australopithecus afarensis*** and was about 20 years old when she

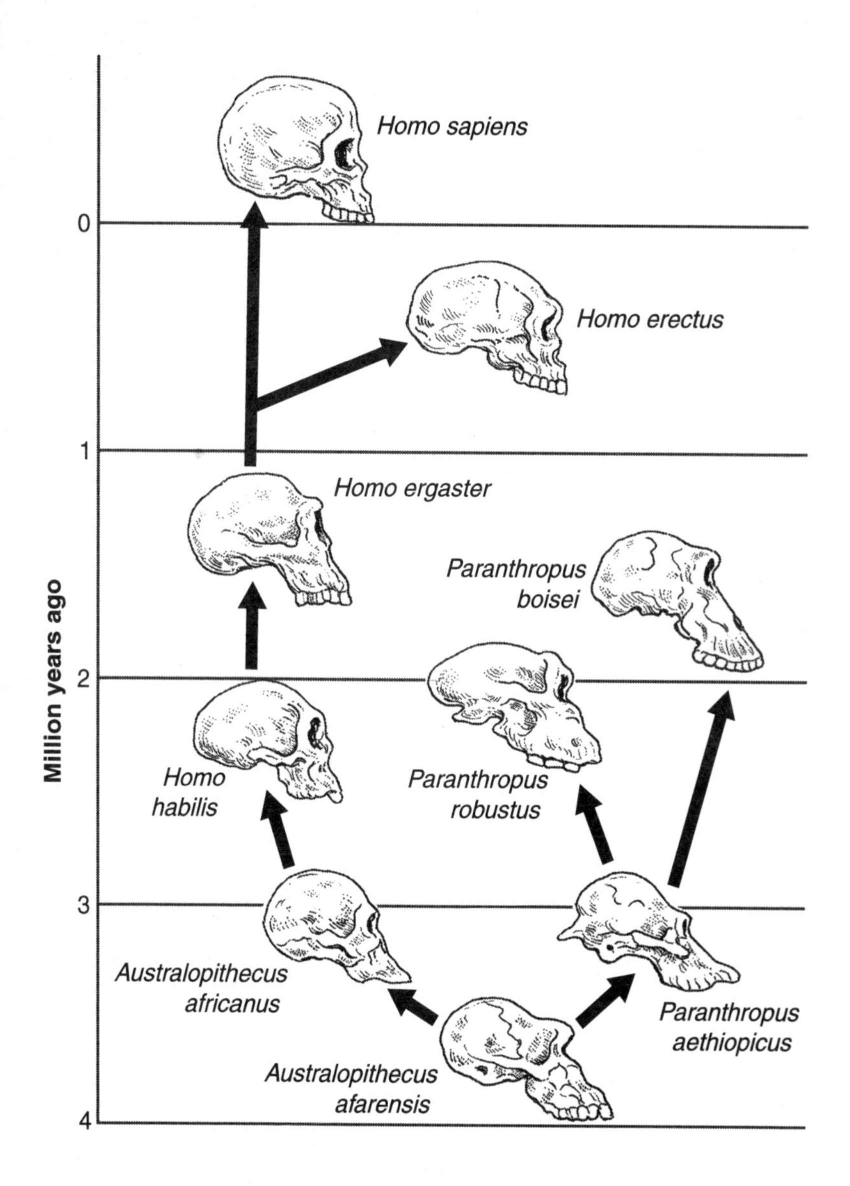

Figure 4.3 The descent of modern humans, *Homo sapiens*

died, perhaps 3 million years before. About 40% of her skeleton was found. She was about 1.25 m (just over 4 ft) tall; males of her species were probably taller. Footprints of this species have been found preserved at the site of Laetoli near the Olduvai Gorge, in Tanzania. Dated at nearly 4 million years old, these footprints were made by three individuals of different sizes, perhaps a male, a female, and a child. The prints formed and solidified in recently settled volcanic ash, along with the prints of primitive elephants, giraffes, three-toed horses, birds and even insects. Until the 1990s, Lucy's claim to fame was that she was the oldest known hominid fossil but in 1994 an older species, ***Ardipithecus ramidus***, was found in 4.4-million-year-old beds in Ethiopia.

After 3 million years, evolutionary changes produced three new kinds of *Australopithecus*. The earliest type, *Australopithecus africanus*, was very similar to 'Lucy' (*A. afarensis*) but it lived in southern Africa and was probably ancestral to our own closest relatives. *A. africanus* also probably gave rise to two other much larger and robust species: *Australopithecus robustus* in southern Africa and *Australopithecus boisei* in East Africa. *A. boisei* had large jaws with huge molars, earning him the name 'Nutcracker Man'. The discovery of this hominid was first made by Louis and Mary Leakey, whose dedicated and meticulous work on fossil people-apes spanned nearly 30 years, from 1931 to 1959. It was Mary Leakey who found the skull of Nutcracker Man and christened it 'Dear Boy' in 1959. Its first scientific name was ***Zinjanthropus boisei*** (*Zing* = 'East Africa'; *anthropus* = 'man'; *boisei* = Charles Bois, sponsor of the expedition that found it).

It is thought that these robust types ate small hard foods such as nuts and seeds, which were plentiful in the expanding savannas of the late Pliocene (about 2 million years ago). It appears that they died out after about a million years without leaving any direct descendants.

The fossil record of about 2.5 million years ago shows the first appearance of stone artefacts made by Palaeolithic (Old Stone Age) hominids. It is likely that they were the first of our own zoological genus, *Homo*. About 500 000 years later, the first definite humans, *H. habilis*, appeared and are known from fossils found in East Africa's Great Rift Valley. Pebble tools associated with this species are evidence of an increasing intellectual capacity and the beginnings of human technology. An increased cranial capacity to about 500–800 cm^3 is evidence of a larger brain and the skeleton was better modified for walking upright.

By about 1.8 million years ago, major changes were under way in the early populations of Africa. A new species, *Homo erectus* ('erect man'), had developed from an ancestor which resembled *H. habilis*, and *Australopithecus* was nearing extinction. Many climatic changes were also occurring, leading to the beginning of the Ice Age. The reasons for the appearance of the larger-bodied and more strongly built *H. erectus* are not clear from the fossil record but it has been assumed that the increased importance of a carnivorous diet was a factor in the development of a new way of life, involving big-game hunting. The brain size of *H. erectus* was 700–1250 cm^3 – just within the range of the brain size of modern humans. A strong, thick-boned skeleton developed in this species which gradually spread throughout the Old World. *H. erectus* spread from Africa to Europe and Asia, first appearing in China about 1.8 million years ago.

A major step in the control of the environment came with the ability to control fire and eventually to produce it wherever it was needed – for cooking, warmth and protection against wild animals. Tools were more sophisticated and featured a range of implements for killing and butchering game.

As the global climate changed and became cooler more than 2 million years ago, ice sheets gradually dominated the landscape. For most of the time, areas such as northern Asia, Europe and North America were covered with ice in the north and there was cold open tundra in the south. While this was happening, semi-deserts and open grasslands spread into tropical and subtropical regions, lakes and rivers dried up and the original homes of early primates shrank. ***Homo sapiens*** ('wise man') succeeded *H. erectus*, making major adaptations to life in the Ice Age.

H. sapiens probably evolved from an advanced form of *H. erectus* between about 100 000 and 750 000 years ago but the transition was gradual, so that in various places there are fossils with characteristics intermediate between the two species. Evidence of the lifestyle of early *H. sapiens* has come from some 300 000-year-old campsites which have been excavated in Germany and France. The oldest preserved wooden artefacts from this time have been found as spears and also as evidence of wooded shelters. In 1998 the oldest known footprints made by modern humans were found in sandstone on the Eastern Cape coast in South Africa. They were estimated to be 117 000 years old, almost twice as old as any that had previously been found.

By about 200 000 years ago a new hominid had evolved in Europe and Western Asia. It became known as **Neanderthal Man**, ***Homo sapiens neanderthalensis***, which successfully adapted to tundra conditions for the first time. For anthropologists, one of the most important cultural developments of these was the habit of burying their dead in caves, with the result that nearly complete skeletons have been found, protected from scavengers and the atmosphere. About 35 000 years ago the Neanderthals began to decline and 5000 years later their place was taken by **Cro-Magnon**, ***Homo sapiens sapiens***. The name Cro-Magnon comes from that of a French cave where artefacts have been found. They appear to have been immigrants from western Asia, where their ancestors lived about 45 000 years ago. By about 20 000 years ago people ancestral to today's 'races' had reached all the major areas of the world, including America and Australasia.

5 HOW EVOLUTION WORKS

The Earth moved

Planet Earth provides the background for the drama of evolution. Just like a stage-play, any change in position or condition of the scenery is bound to have an effect on the scene being acted. As an analogy, geographical changes have been part of scenic changes which have been the most important selective pressures to have shaped the evolutionary destiny of all living things. The very surface of the Earth is dynamic. When it moves only the best adapted to change are able to survive. Extinction is the eventual fate of failure to adapt.

The changeable nature of Earth is seen in the unpredictable patterns of earthquakes and the spasmodic eruption of volcanoes. However, these are but temporary, if spectacular, proof of the massive forces that constantly move the continents across the surface of our planet. Evidence for such movements began to accumulate as long ago as the seventeenth century, when the first maps of the southern hemisphere were made. **Francis Bacon**, the eminent philosopher and scientist, became curious about the apparent match between the coasts of Africa and South America. The science of geology was non-existent then and explanations of such observations did not materialize until hundreds of years later.

It was **Alfred Wegener**, born in Berlin in 1880, who proposed the idea of continental movements as long ago as 1912. Like Bacon, Wegener noticed that the mapped outlines of Africa and South America fitted together like two pieces of a jigsaw rather neatly – and he did not dismiss it as a coincidence. A closer look suggested to him that by joining the continents together, it was easy to explain how geographical features appear to continue across now distant land masses. His hypothesis could also account for the similarities in the distribution of many plants and animals on both sides of the Atlantic, so he proposed that the continents had drifted apart.

Today the idea of **continental drift** is accepted by all serious evolutionary biologists and geologists but when Wegener published *The Origin of Continents and Oceans* in 1915 the idea of floating continents on a sea of rock seemed absurd. Land bridges, which had disappeared long ago, were thought to have allowed creatures to have crossed from one continent to another, thus explaining the similarities between some species that are now found on opposite sides of oceans. It was thought that mountain ranges had formed due to the cooling, shrinkage, and consequent wrinkling of the Earth's crust rather than being the results of continental collisions. Wegener's theory was ridiculed by most of his contemporaries. One American geologist of the time is quoted thus: 'discussion of it merely encumbers the literature and befogs the mind'; another described the idea as a 'state of auto-intoxication'. The paths towards many original and innovative hypotheses are often littered with obstacles. As a result of Wegener's suggestion no German university would employ him and so he moved to Austria to continue his teaching.

Like that of many brave new thinkers before him, his work was not recognized until many years after his death – it was not until 1958 that geological opinion moved towards acceptance of Wegener's theory. New techniques for measuring the effects of magnetism revived interest in continental drift but some of the mechanisms that people suggested to explain the force necessary to move continental masses now seem strange. They included the idea that vast masses of polar ice set the Earth's crust slipping around its core. Soon Wegener's explanation seemed more plausible and it was the end of the contrary opinions of his opponents.

The most supportive data for continental drift was based on magnetism in ancient lava flows. When a lava flow cools, metallic elements in the lava are oriented in a way that provide permanent evidence of the direction of the Earth's magnetic field at the time, recording for future geologists both its north–south orientation and its latitude. From such maps it is possible to determine the ancient

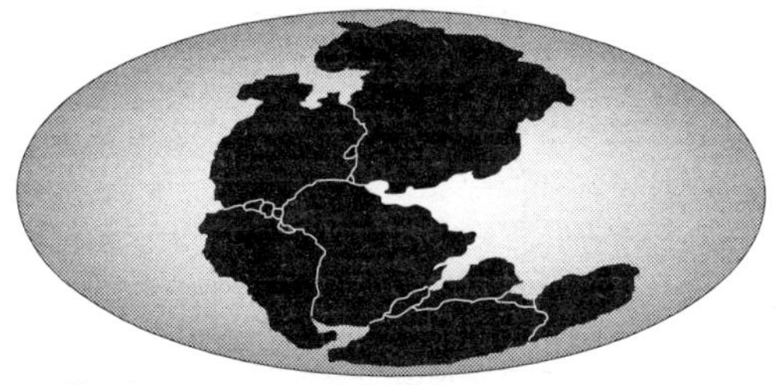

One big continent 200 million years ago in the early Mesozoic era–Gondwana

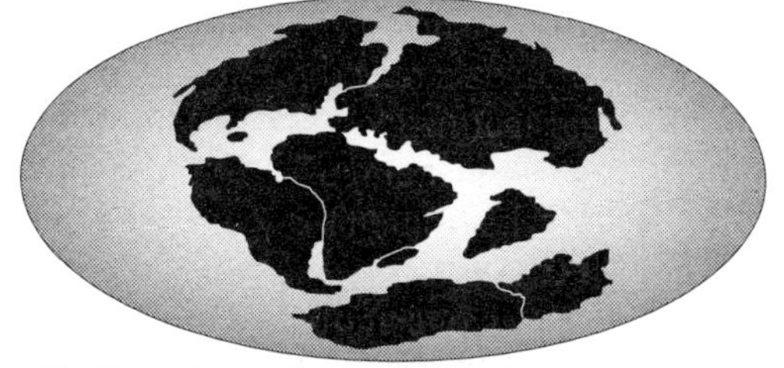

The big continent split up towards the end of the Jurassic period, 135 million years ago

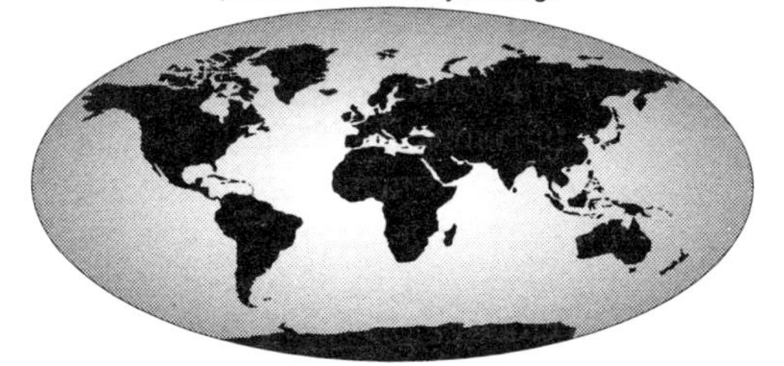

When and where the continents drifted to make today's world

Figure 5.1 Continental drift

positions of today's continents. It is now believed that not only has continental drift occurred but that it continues to occur today. Astoundingly precise measurements made by satellites have shown that continental drift, or **plate tectonics**, is resulting in an annual 5 cm increase in the width of the Atlantic Ocean!

Wegener's work had revolutionized geology and an institute was set up and named after him in his honour. Ironically, he was not there to enjoy his recognition. In 1930, having embarked on his third expedition to Greenland, he perished while returning from an heroic cross-country trek to resupply a winter camp on a high plateau. His body was found a year later, frozen in his sleeping bag beneath the snow.

Wegener's epoch-making breakthrough in geology has shown the great evolutionary significance of the Earth's movements on a grand scale. South America is an example of the great influence of continental drift on evolution. Combined with Africa, India, Australia and Antarctica, it was once part of a single continent now named **Gondwana**. The area we now call South America broke away about 100 million years ago. Antarctica became separated from it about 30 million years ago, leaving South America isolated as an island continent, separated from North America throughout most of the Age of the Mammals.

South American mammals consisted of marsupials and a group of primitive hoofed types resembling horses, rhinoceroses, rodents and bears. Through **convergent evolution** they developed forms that were superficially similar to those that evolved under similar conditions in other parts of the world. For example, on the plains there were single-toed running grazers which looked like the small horses of North America but which were in no way related. The woodlands had huge browsers which resembled the elephants in other parts of the world.

The South American fauna existed until about two million years ago, when at the close of the Pleistocene a land bridge, formed by volcanic action, joined North and South America. This allowed migration between the two land masses. Most of the movement seems to have been one way, from north to south. Gradually the northern species replaced the southern ones, except for a relatively few species such as the marsupial opossum which still exists as the Virginia opossum.

The success of the northern species over the southern species is a good illustration of how competition leads to greater potential for adaptation. The almost complete replacement of southern types by competitors from the north was a consequence of intermixing between North American animals with those of Asia. Competition resulted in the northern types being under constant selective pressure from new and varied Asiatic species. They therefore evolved into more adaptable species. In contrast, the southern types had remained virtually unchanged throughout the time of the continent's isolation because there was no competition from outside. They had therefore become highly specialized and unable to meet the competition from the more adaptable animals which had migrated there from the north.

Like the mammals of South America, the Australian marsupials evolved and thrived because of almost total lack of competition from the 'higher' placental mammals. This happened because Australia broke away from Gondwana before most placentals appeared in that area. Australia originated adjacent to Antarctica and drifted northwards until it reached the desert and tropical grassland belt of the southern hemisphere. The fauna would have had to adapt constantly to accommodate this change. In the next 50 million years or so, Australia will continue to drift northwards into the tropical forest zone, and many more marsupial forest animals will have to adapt to survive – providing they escape extinction as a result of human activity. By that time the continent may well have collided with Asia, allowing a mixing of the fauna between both continental masses.

This highly theoretical conjecture is obviously being over-ridden by human influence on the native fauna of both Asia and Australia. In the last few hundred years the introduction of sheep, rabbits, camels, cats, etc. to Australasia has threatened almost all marsupial species and it remains to be seen how they will cope with these competitors together with the destruction of their habitats.

Marsupial fossils have been found in Antarctica, apparently proving the former link between South America, Antarctica and Australia. However, it is among the reptiles of the Permian that the classical evidence of the former existence of Gondwana is found. Fossils of the same species of reptiles have been found in rocks of Brazil and in South Africa. Indeed, rocks in Africa, India and Antarctica have been found to contain the same species of reptilian fossils. The contemporary plant life was also the same on these, now widely separated, land masses. Along with certain geophysical evidence, such fossil evidence is regarded as the classic proof of Wegener's theory of continental drift.

One of the effects of the final break-up of Gondwana 50 million years ago was the separation of India from Africa. At the same time many other pieces of eastern Africa sheared off and are now found spread throughout the Indian Ocean. These include the Seychelles and Madagascar. The indications are that the splitting process has not yet finished, and East Africa itself may divide along the Great Rift Valley and move into the ocean.

Madagascar is an important example of an island isolated in this way. Its zoological significance lies in the fact that it is the home of the lemurs. These are primitive relatives of the monkeys and apes. Madagascar broke away from the African continent when these primitive primates were widespread and successful. They continued to survive on the island in the absence of competition and now they occupy a large number of ecological niches there. The **lemurs** of Madagascar are unique because of their special adaptations.

Some places on the Earth's surface, such as volcanic islands, have only recently been inhabited by land-living animals and plants, and without human interference.

Volcanic islands can form as a result of the creation of new crustal material at oceanic ridges or as a side-effect of the destruction of the crust in deep ocean trenches. After a period of cooling, the arrival of living things tends to follow a particular pattern. The first arrivals are seeds and spores, especially those with adaptations like the parachutes of dandelions which are blown there by the wind from distant populations. They root in the cooling mineral-rich ash. The first animals to parachute in are tiny mites and spiders, because of their lightness. Flying insects, helped by the wind, usually arrive next. Birds soon arrive and provide a rich diversity when they are able to exploit plants and invertebrates as food. Colonization by ground-dwelling animals is a slower and more haphazard affair. After many years, lizards, tortoises and rodents may land on the new island, having drifted out to sea from distant land masses on uprooted trees or other vegetation at times of flooding.

In the absence of land-dwelling predators on islands, some birds can survive even if a series of mutations prevents them developing functional wings. New Zealand was once an example of this. In an island group that possessed no mammals except bats and seals, giant flightless birds like moas were the largest living animals. Their only surviving relative is the kiwi although many other flightless birds retain their wings but use them only for balance while moving. In other isolated island habitats, a single species may rapidly radiate to occupy all the available niches.

The cost of this sort of specialization is vulnerability to exploitation. The introduction of novel selective pressures such as the arrival of humans with their associated livestock and pets usually sounds the death knell for such evolutionary naive species. Goats and sheep quickly compete for vegetation. Rats eat eggs and anything else that cannot escape them. This form of destruction has happened many times throughout the world. Perhaps the most well known is the extinction of the dodo of Mauritius.

Another aspect of the movement of the continents that has influenced evolution profoundly during the past few million years is the continental configuration near the north and south poles. Throughout most of Earth's history the positions of the continents have been such as to allow the free circulation of sea water around the globe. This has had the effect of transferring heat from the hot areas to the cold ones, and so an equable climate was maintained over most of the Earth. During the Age of the Mammals, the continent of Antarctica settled firmly over the south pole. This meant that the interior of one of the coldest areas on Earth was denied access to the warming effect of ocean currents, and so a permanent ice-cap developed upon it.

Similarly, the continents of the northern hemisphere grouped themselves in a ring around the north pole, producing an almost land-locked ocean. No warming currents meant that the Arctic Ocean froze to become another permanent ice cap. The last few million years probably represent the only time in the Earth's history when the planet has had two ice caps at the same time.

During the last Ice Age so much water was locked up in ice sheets that the ocean level was lowered, which created bridges and opened migration routes between continents, such as between Britain and the rest of Europe. As a result, different animals could mix and compete, which caused some species to die out and new ones to evolve.

The cold conditions produced on the northern continents were the selective pressures that allowed mammoths, mastodons and woolly rhinoceroses to survive. They probably also contributed to the evolution of species that developed organizing skills and intelligence to survive ever-changing harsh conditions. Members of this species, on the primate line of evolution, gradually lost their body hair as they learned to construct shelters, use fire and eventually to make clothes. They became the proud inheritors of the title *Homo sapiens* – 'wise man'.

The beckoning finger

We saw earlier (page 12) that Charles Darwin recognized the importance of variation as he uncovered its role in the process of evolution. Variation, he implied, provides the raw material upon which natural selection can act. Some of the members of a population will possess inherited traits that enhance reproduction. Those traits will, therefore, be carried into future generations. Furthermore, as they are carried along they will replace less successful alternative traits, and so they will come to increase proportionately in later generations. Natural selection becomes a finger beckoning to the otherwise unguided variations in animals and plants: all other principles and facts of evolution may be logically related to it or explained by it. The best part of a century and a half has passed since Darwin's masterpiece *The Origin of Species*, but his idea remains triumphant. It might have been a misunderstanding, after not having read and understood Darwin's work, or it could have been an attempt of the media to sensationalize but it was unfortunate that a nineteenth-century journalist coined the term **'survival of the fittest'**. As a result, 'fittest' became to be used only in the context of 'strongest' but actually, in the objective way that evolutionists use it, they mean the 'fittest to breed'. Thus, some drab, insignificant-looking animals that are more efficient breeders than some of the more spectacular looking types have a greater fitness. In evolutionary terms then, we can expect the greatest success from the best reproducers. Thus each generation will be composed primarily of the offspring of the best reproducers from previous generations, and we can expect those traits that led to reproductive success to increase in frequency from one generation to the next. Nevertheless, variation will remain in the population for virtually any trait, including those directly related to fitness.

Variation for any trait in a population can be shown graphically as a Normal distribution (Figure 5.2) Imagine that this curve represents height in a population of humans of the same age and gender living in a particular country. We see that most individuals in the population are of intermediate height, and that there are increasingly fewer taller or shorter people deviating from that particular height. Any trait distributed in this way in a population produces this Normal, bell-shaped, curve but of course the shape can change according to circumstances. For example, the curve may be higher and narrower; it can widen; it can be skewed to the right or left, or it can be bumpy. What do these shapes mean in the context of evolution?

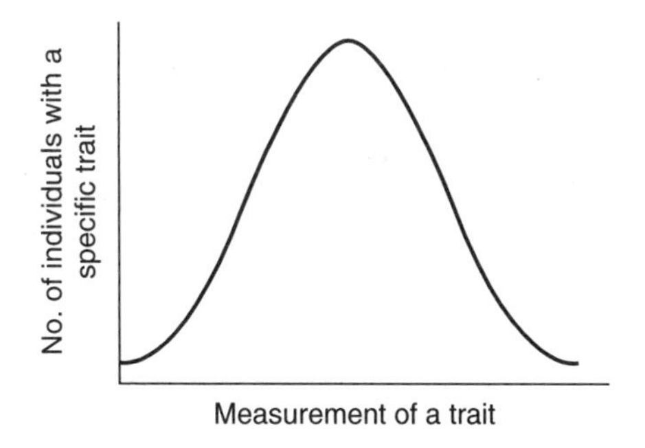

Figure 5.2 Normal distribution. A statistical term that is graphically depicted as a bell-shaped curve

First, if being of intermediate height is especially important, then those people who are taller or shorter will suffer some disadvantage. This is called **high selection pressure** (strong selection against departure from the optimum). In time, there will be fewer of the extremes and the curve will become narrower, as we see in Figure 5.3(top).

If height is not so important, shorter and taller people will tend to survive and reproduce and, because of **low selection pressure** (weak selection against departure from the optimum), the population will produce a wider curve (Figure 5.3(bottom)). So when the environment is stable over a long period of time, as it is assumed in these cases, the population tends to cluster around a single type, with divergent forms being less successful to some extent. The result is **stabilizing selection**. The mean or average condition is assumed to be optimal for the prevailing conditions if stabilizing selection is being demonstrated.

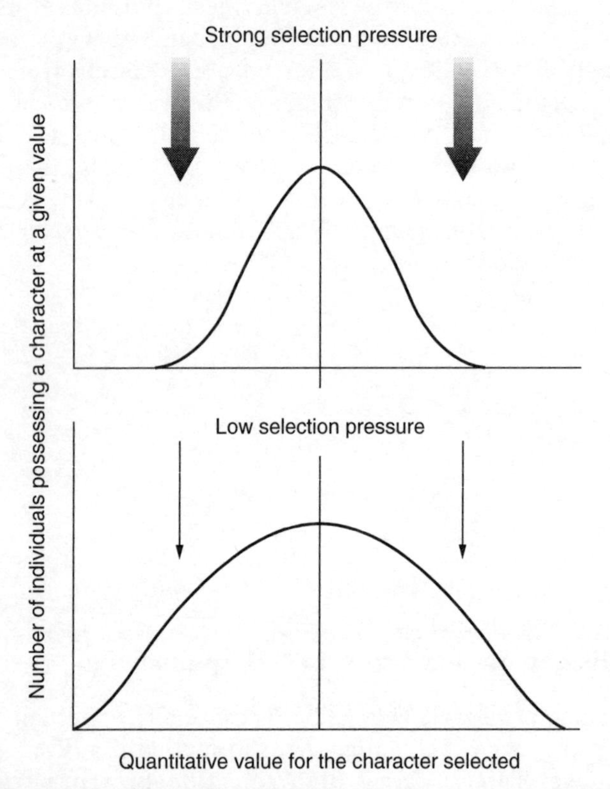

Figure 5.3 The influence of selection pressures on population. See text for details

Yet conditions can change. Suppose that taller becomes an advantage – remember how giraffes evolved. The population then shifts toward the high end of the height scale, as in Figure 5.4.

Such a process is called **directional selection** (change due to populations moving towards a new optimum for a particular trait).

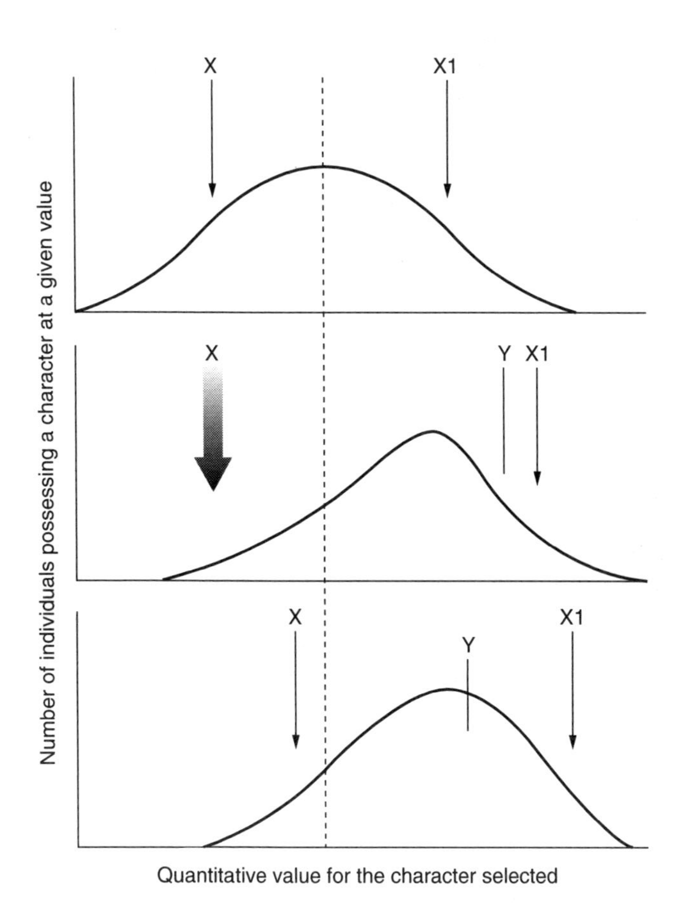

Figure 5.4 Directional selection. Y marks the optimum value for some trait, such as height. Those taller or shorter would therefore not do as well. Their numbers would be reduced by ecological pressures (shown by X and X1). If it should become better to be taller then Y would shift. Individuals at X would suffer while those at X1 would thrive. The result would be a taller population until the population stabilized around the new optimum

In some cases there may be selection against the mean condition, resulting in an accumulation of individuals at the extremes. This is called **disruptive selection**. For example, consider a population of

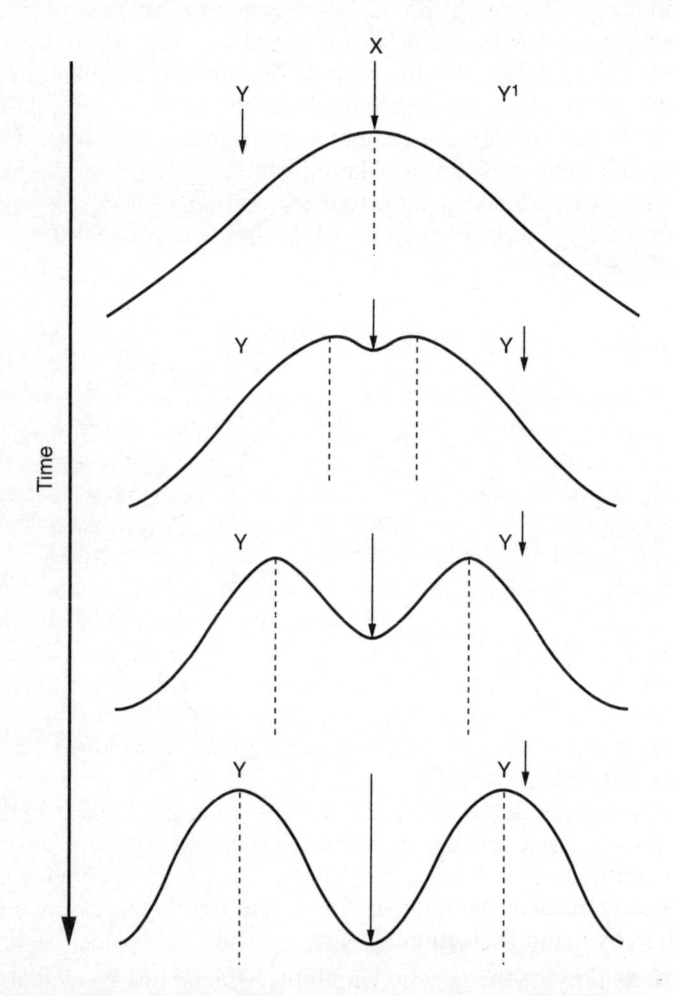

Figure 5.5 Disruptive selection can produce a bimodal (two-humped) curve. Bimodal curves are produced in populations in which it is advantageous to be at one of two extremes of a curve since the intermediate condition is selected against

marine skates. They normally escape predators either by lying flat on the sea bed covered with sand or by rapidly moving away. Suppose most of the skates try concealment for a time and then begin to flee when a shark approaches. Others though, do not move, remaining hidden as the shark swims over them. Yet others flee at the first sign of danger. In this population, then, we might expect selection against the average behaviour as the sharks caught those that tried to flee while the predators were near. If others tended to escape more frequently, the population would tend to be composed increasingly of the hiders and the runners. If escape tendencies were plotted on a graph, the population would produce a **bimodal** (two-humped) curve, as we see in Figure 5.5.

Raw materials for variation

Evolution can be large scale or small scale, the latter being frequently overlooked. A species always shows variation between its members and so may respond quickly to selective pressures, by shuffling and sorting the genes it already possesses – otherwise dog breeders, with their deliberate artificial selection, would never have been able to produce the vast variety of familiar breeds we see today, some of which have been developed in a matter of years. The pool of variety within the genes of a species is a sort of adaptation in itself, and a most important safety valve in natural selection in ensuring survival if conditions change. There is such a thing as being too well adapted to one specialism. If the species cannot virtually jump to the crack of a selection whip and adjust itself quickly to slight changes in its environment, it will fail Nature's ultimate challenge and become extinct. The geological record in the rocks show us that this has happened very often. If good fortune smiles on the species, it may be able to produce a new combination of its genes to meet the new situation, and this is the essence of evolution by natural selection.

Mutations provide the genetic variability which must be available before selection can take place. In one sense mutations are no more than misprints in the **genetic code**. The replication of genes is an extraordinarily complex process which occurs during cell division. It is not surprising, therefore, that errors sometimes lead to misreplication or some other disruption of the normal process (the 'spontaneous mutations'). Mutations recur at a very low frequency, at a constant rate, and are random in their effect with respect to the environment. There are local concentrations of natural environmental factors which tend to increase them, such as ionizing radiations and certain groups of chemicals called **mutagens**. The mutant **allele** may be dominant, intermediate, or recessive in effect, but the most common classes are recessive or partially recessive. As a rule the mutation is deleterious. If severely disadvantageous and dominant, the mutation is exposed to selective pressures as they arise, but if the mutation is recessive, it will be subjected to selection only when **homozygous**. Occasionally a mutant form happens to possess characteristics which fit it for survival in a newly changed environment. It is said to be **preadapted** to the new environment, and it increases in frequency.

In experimental conditions, the production of new mutations by artificial means, especially by irradiation, is sometimes used to aid selection programmes. Mutations which might have occurred spontaneously are induced at a higher frequency, making them sufficiently common to be found in quite small samples. They are then retained by selective breeding and used in subsequent experiments. Such work is a model of the natural process of evolution in which the selective pressures and mutation rates are greatly magnified.

In the 1950s much concern was expressed about the effects of radioactive fallout at a time when nuclear bombs were being tested by the 'super powers'. **H.J. Muller** published a famous paper, in which he described the fraction of the population dying as a result of mutation as the **mutational load** and questioned whether populations, especially human, could tolerate a raising of the load they had to bear. The rate of mutation of individual genes induced by radiation is linearly related to dose. A doubling of the radiation above background level would double the mutational load. Muller's concern was subsequently discussed at great length – and by the 1990s, as a result of using more sophisticated techniques to study the effects of the Chernobyl disaster, hard data became available on the viability of populations in relation to deleterious mutations.

In strictly biological terms, looking at animals and plants in general, it is possible that they can tolerate this load, which is manifested in a variety of ways from death of the individual to a reduced reproductive rate. Survival is often controlled by ecological or social factors, so that loss of one offspring carrying a deleterious mutation is followed by production of a replacement. Several genes often act together to control the same trait and the individual is adversely affected only when more than one of these genes is affected. These, and other, considerations suggest that mutational load is not an important threat to the continued survival of most species. On balance, it seems probable that most plants and animals have no difficulty in supporting an increased mutation rate resulting from average levels of radiation as atmospheric pollution. For example, much detailed research into the wildlife in a 30-km radius of the Chernobyl disaster has taken place since the accident happened in 1986. By 1995, the extent of genetic mutation was clearly seen; no fewer than 46 mutations in just one gene were found in nine voles collected within the area studied. There were only four mutations of the same gene in samples taken from outside the area, yet the seemingly high mutation rate did not eliminate the vole population.

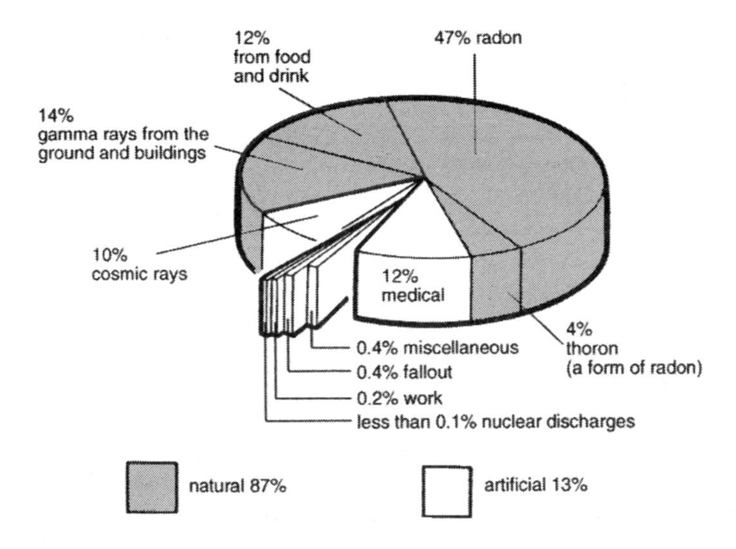

Figure 5.6 Sources of radiation in the environment

Humans, however, present a different situation. This is because people are protected from the action of natural selection. In modern society, men suffering from haemophilia have a much better chance of surviving to pass on the defective gene to their offspring than they would have under prehistoric conditions, so the statistical estimate of the frequency of this mutation may be raised. Some sources of radiation in our environment are shown in Figure 5.6.

Variations under pressure

In the 1700s, poets extolled the virtues of the green British countryside at great lengths and anyone who reads Gilbert White's *Natural History and Antiquities of Selborne* of 1789 cannot fail to appreciate the descriptions of the rustic scenery of the time.

> … The pendant forests, and the mountain greens
> Strike with delight; there spreads the distant view,
> That gradual fades till sunk in misty blue:
> Here nature hangs her slopy wood to sight,
> Rills purl between and dart a quivering light …
>
> From *The Invitation to Selborne*, Gilbert White, *circa* 1763

The trees were numerous then and many were clad in light grey lichens. Furthermore, if one looked carefully enough at the lichen coverings, it was sometimes possible to see a camouflaged light-coloured moth sitting concealed upon the bark. One might occasionally spot a black moth resting rather conspicuously against the light-coloured background. The black moths were rare, largely because they were easily seen by hungry birds and quickly removed from the population. 'Progress' eventually had its way and Britain's entry into the industrial revolution was to change its 'green and pleasant' land for ever. It was marked by the construction of huge coal-burning factories which spewed out their effluent as black sooty smoke. Soon the green countryside so beloved by Gilbert White and his contemporaries quietly succumbed to a cloak of soot from the 'dark satanic mills'. As the tree bark and lichen darkened, the more common light-coloured moths became more visible to bird predators and the dark coloured moths became progressively camouflaged.

This was **industrial melanism** in action. The increase in frequency of black forms of this peppered moth, *Biston betularia*, is probably the most famous example of genetic response to selection imposed by pollution. The phenomenon was first observed in Manchester, which had been the centre of an area of smoke pollution since the beginning of the nineteenth century. The black (melanic) form of the peppered moth, called *carbonaria*, was first reported by collectors there in the late 1840s; by 1900 it had almost replaced the light-coloured typical form. The increase from around 1% to 99% took 50 generations (the moth has an annual life cycle). During the second half of the nineteenth century the melanic form began to be reported from other parts of Britain as a result of emigration from the centres of population increase. The black form was first recorded in London in 1898 and rapidly ousted the typical form by becoming between 80% and 90% of the population. It reached Belfast by 1894, and increased to about 75% even in the comparatively unpolluted atmosphere of Norfolk. By the 1970s its numbers were decreasing.

Is the peppered moth an exception to other moths? The answer is certainly 'No'. Over 150 species of moths have been proved to exhibit industrial melanism, but the peppered moth is one of the most striking examples and was meticulously documented in *Selection experiments on industrial melanism in the Lepidoptera* by **H.B.D. Kettlewell**. It seems very likely that there are hundreds of other examples world-wide. Evidently a threshold level of pollution was required before natural selection occurred. London certainly suffered from coal smoke pollution as far back as the seventeenth century, yet significant numbers of melanics appear to have reached there initially by immigration from other centres.

The greater conspicuousness of the typical forms and the greater protection of the melanics in industrial areas is due not only to blackening surfaces by soot but also to the destruction of lichens growing on tree bark. These are very sensitive to another air pollutant from burning coal – sulphur dioxide. The Clean Air Act of 1956 resulted in a vast reduction of coal burning. Consequently, there was a re-colonization of niches previously occupied by many species of lichen. At the height of atmospheric pollution by sulphur dioxide, monocultures of the sulphur-tolerant lichen *Lecanora conizaeoides* developed in areas subsequently laid bare. This lichen increased during the pollution, having been rare in Britain previously. The result was an increase in the typical form of the peppered moth, which was no longer at a disadvantage. Populations reverted to the condition previously noted in the days of Gilbert White. Yet another one of nature's cycles has been completed.

Another example of an environmental selective pressure caused by pollution, and influencing the evolution of populations, is seen in tolerance to heavy metals by plants. A number of grasses show genetic adaptation to high concentrations of metal ions in areas where spoil heaps from mines have brought potentially toxic heavy metals to the surface. In Britain many deposits of waste from zinc and lead mining are over 100 years old but are still devoid of plants; nevertheless, some species of the grasses *Agrostis* and *Festuca* can colonize them. Plants from such sites show tolerance of the metal when grown in experimental conditions and since the 1970s more than 21 species of flowering plants have been identified as being heavy metal tolerant throughout the world. The tolerance is inherited and controlled by several genes.

Selection for heavy metal tolerance has thus led to genetic response in some species, allowing them to colonize newly available niches. Tolerant individuals of *Agrostis* occur at a frequency of about one or two per thousand in populations not exposed to toxic metals and there is evidence to suggest that the successful growth of tolerant plants on non-toxic soils is adversely affected by competition with non-tolerant plants, so that the former tend to be eliminated. The disadvantage may not be very great, however, so that tolerant individuals persist in populations. The reverse is certainly not true on the spoil heaps. In such situations, non-tolerant plants are non-viable and only tolerant plants survive. The result is that patches of metal-tolerant plants may be of very small size, and they show a sharp boundary at the edge of the toxic sites. Some zinc-tolerant individuals have been found in belts no more than 30 cm wide beneath zinc-coated metal fences. When a new toxic spoil heap is created containing high levels of zinc, copper, or lead, rapid colonization by certain species of plants can be expected, starting from tolerant individuals present in normal populations. Since the selection for tolerance can be extremely strong, a population can change from almost entirely non-tolerant individuals to only the few tolerant survivors within a single growing season.

New bugs for old

We use a number of chemicals, such as pesticides and antibiotics, to protect us from competitors for food and from pathogens. However, the substances that we use can become pollutants when they exert effects in ways other than controlling their target organisms. The most notorious example of this began with the first insecticide to be used on a world-wide scale. In 1935, a Swiss chemist, **Paul Muller**, began a search for a chemical that could interfere with some of the biochemical reactions that go on in insects. In September 1939, he came across a compound called dichlorodiphenyltrichlorethane. The name is even too long for biochemists to use, so today we all know it as **DDT**. The compound was first prepared and described as long ago as 1874 but at that time it did not appear to be very special. Muller, however, found that it was just the thing he was looking for. It was cheap, stable, odourless, fairly harmless to most forms of life, but deadly to insects. During the Second World War large amounts of it were used to combat a major typhus outbreak in Naples during its occupation by the Allies in 1943. The causative agent for typhus is carried by insects such as lice. The use of DDT spread throughout the world and many regions that are now free of killer diseases such as malaria owe their freedom from insect vectors to DDT. Yet this does not represent a happy ending.

The gene pool of insects with such large and prolific populations is better described as a gene ocean. There is always the odd mutant with a slight variation in genetic make-up wherever there are such vast numbers. Maybe one in a million will have dormant genes to control the production of an antidote to a chemical insecticide. That will be enough! It was certainly enough to build up populations resistant to DDT. The presence of the insecticide is a selective pressure which works to the resistant insects' advantage. They multiply and then a whole new type of resistant insect comes on the scene. Thus, as years passed by, DDT became less effective on the housefly, for instance. Some resistance was reported as early as 1947. Today, every insect which was originally attacked by DDT has resistant forms. With all of human technology we have not managed to eliminate even one species of unwanted insect.

The widespread use of insecticides has provided such sudden alterations in some insects' environments that fundamental adaptive changes have occurred within a few years. It has led to levels of resistance which make the insecticide impotent. Once a population has evolved resistance, further applications of the increasingly ineffective insecticide are no longer a control measure, but merely pollute because the chemicals can build up in food chains, reaching harmful levels in the creatures at the top. Every year reports appear of fresh instances of resistance, and doubtless many examples are never detected because they are of no social or economic importance. By the end of the 1960s over 225 insecticide-resistant species had been officially recognized, but this probably represents a fraction of the true figure. A large number are resistant to the cyclodiene group of chemicals (dieldrin, aldrin, lindane etc); next in importance is the DDT group (DDT, DDD, methoxychlor etc); while the organophosphates (malathion, fenthion, etc) are third in importance. Twenty species are resistant to other compounds which are in none of the three major groups.

Many species have become resistant to insecticides of different groups. Probably the most striking example of such multiple resistance is found in the housefly, *Musca domestica*, which, in some parts of the world, is now resistant to almost every insecticide which can safely be used.

Resistance arises as a result of selection acting upon natural variation and is an example of evolution in action. There are several sources of evidence demonstrating this but some of the most enlightening date from the 1960s when research on the fruit fly, *Drosophila melanogaster*, by geneticists involved selectively breeding the flies which showed natural resistance to DDT. The life cycle of this species is about three weeks, and after several generations artificial selection produced completely resistant populations. Similarly, after several generations, mutations for resistance to other insecticides have been artificially induced in *Drosophila* by irradiation. Doubtless such mutations occur naturally in wild populations from time to time, although at much lower frequencies.

Some insecticides act as gene switches in insects, causing genes to act to control the production of certain enzymes which can break them down. Resistance may be dominant, recessive or intermediate in first-generation offspring and within a given population it often appears to be under the control of a single gene. This is the case with resistance to dieldrin. However, at least three different interacting genes cause resistance to DDT in the housefly.

While some species evolve resistance rapidly, others seem to lack the potential altogether, despite many years of exposure. However, once resistance arises it can spread rapidly as evidenced by the malaria-carrying mosquito in India. DDT was used for over 10 years before the first resistant strains were reported in 1959. By 1965 they were reported from widely separated Indian states, and today resistance is seen in mosquitoes from most of India, Burma, Sri Lanka, Pakistan and Iran.

If this seems like rapid evolution in progress, it is very slow compared with the rate at which bacteria evolve. Antibiotic resistance in bacterial pathogens poses serious health problems particularly in parts of the world where antibiotics can be freely purchased from pharmacies. The typhoid epidemic of 1972 in Mexico was an infamous reminder of this. Under tragic circumstances there was found to be widespread resistance by the typhoid bacillus to chloramphenicol. As an evolutionary phenomenon, antibiotic resistance has many parallels with resistance to insecticides and to other drugs such as those used against the malarial parasite itself. Antibiotic resistance arises basically by the selection of preadapted resistant mutants. Since the early 1950s it has been known that it is possible, as with insecticides, to select for resistance and to produce pure cultures of resistant strains which have never been in contact with the antibiotic.

The genes and cellular mechanisms involved in resistance vary between the different antibiotics and between species, and this is reflected in the rate at which resistance can be selected for. However, even within a species the mechanism of resistance to a given compound may be variable. Penicillin resistance has been recognized since the 1940s and arises because of the action of a bacterial enzyme which can break down penicillin (most naturally occurring penicillin-resistant strains of bacteria owe their resistance to this enzyme). Resistance to antibiotics has already destroyed or limited the usefulness of several formerly valuable drugs and selection of resistant bacteria in the presence of the drug can be rapid.

Superimposed on this problem is another which seems to be unique to bacteria. Some strains can transmit their genes controlling resistance to others during a process of **conjugation** among sexually active individuals. This discovery was first made in 1959 by Japanese scientists who demonstrated that the genes are transferred on strands of DNA like a rope being passed between two 'mating' strains of bacteria. This discovery of **transferred resistance** caused a reappraisal of the use of antibiotics in animal husbandry, where they were employed not only to cure infectious diseases but also in attempts to prevent infections, and in some cases as food additives to promote the growth of animals. The urgency of the situation became clear in Britain in the mid-1960s when an epidemic in cattle of *Salmonella* was shown to be resistant to seven well established antibiotics. Farmers and vets caught the infection and six people died, despite treatment with antibiotics. The harmless gut bacterium, *Escherichia coli*, of these cattle were also highly resistant and 99 of the 134 multiple-resistant strains found were also able to transfer their resistance to sensitive strains of *Salmonella* and *E. coli*. The seriousness of the situation led to the setting up of the Swann Committee in 1975 to consider the problem. It recommended that no antibiotic with a therapeutic value in relation to human disease should be used to promote growth or as a preventative measure in animals, although such drugs could still be used for the treatment of animal disease.

Equally likely to promote resistance is the uncontrolled medical use of these drugs, either through unrestricted sale or through over-prescription by doctors. Indeed, it is now common practice to use antibiotics as the last resort only when the body's natural resources have been shown to be inadequate.

In conclusion to this consideration of evolution in action, we must remember that our activities are affecting natural selection. Just as insects and bacteria evolve resistance, so do rats to rodenticides, plants to herbicides, malarial parasites to anti-malarial drugs, and plants and animals growing on the hulls of ships to antifouling paints. Response to the new selection pressures often occurs rapidly. In many cases, such as the rise of heavy metal tolerance in plants, natural selection serves to compensate for the polluting effects of human activities. In others, such as pesticide resistance, the result is to decrease the cost-effectiveness of the application to the target species and to increase general environmental pollution if more concentrated doses come to be used.

So far we have looked at the effects of artificially changing the environment of some of the most numerous species on Earth. Insects and bacteria can be counted in millions of millions throughout the world. Let us remember a fundamental biological rule:

Given a sufficiently large population and sufficient time, species will adapt to their surroundings in such a way that it is to their advantage.

We certainly see this in the examples discussed above, but what is the effect of selective pressure on smaller populations?

Genetic drift

Small populations may be moulded by different factors than those operating in larger populations. For example, in small populations **genetic drift** may be important. Genetic drift is a change in the frequency of genes in a population due to simple chance. Changes of this sort are not produced by natural selection. Genetic drift is important in small populations because they would, by definition, have a smaller **gene pool** (reservoir of genes). This means that any random appearance or disappearance of a gene would have a relatively large impact on gene frequencies. In large populations, any random change would have little effect since it would be swamped by the sheer numbers of other genes in the population. There are two important ways that small populations can set the stage for such random events.

The bottleneck effect

Certain kinds of disasters can reduce a population unselectively – that is, without favouring one genetic type over another. Among these are earthquakes, floods, fire and drought. The result can be a

genetic bottleneck and occurs when a population is drastically reduced, but the remaining population is not genetically representative of the original population. By simple chance, some alleles will be over-represented and others under-represented (Figure 5.7)

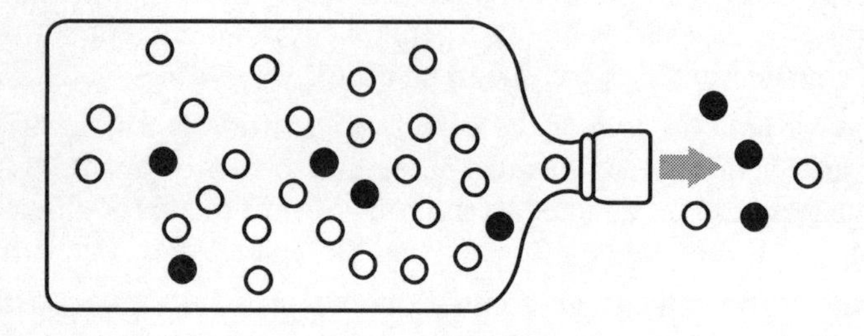

Figure 5.7 The bottleneck effect

For example, a large population (inside the bottle in the diagram) may have a certain gene frequency (say light to dark individuals), but a small part of that population (outside the bottle) may have a different gene frequency. If this smaller population then serves as the breeding stock for the next generation, that generation will have a different gene frequency from the original population.

Populations of elephant seals and cheetahs are each genetically remarkably similar, since it is believed that both groups have recovered from very low populations at one time.

The founder effect

Another influence operating on small populations is called the **founder effect**. Occasionally, a few organisms will be separated from the main population and establish their own breeding group. For example, a few birds might be blown from the mainland to some remote island, as probably happened in the case of Darwin's finches (see page 14). These became the founders of a new population. Sometimes storms might wrench a large tree loose, allowing it to float out to sea. Such a raft could carry small creatures and eventually strand them on some new land, perhaps an island (Figure 5.8).

Figure 5.8 The colonization of islands

Evidence of this method of transport was obtained in 1995 when Ellen Censky of the Carnegie Museum of Natural History in Pennsylvania reported that fifteen green iguanas rafted on a mat of vegetation to Anguilla from Guadeloupe, a journey of about 300 km. This followed Hurricane Louis when a hillside was washed into the sea. The iguanas arrived on Anguilla alive, and one healthy female was sighted there 29 months later.

Insular implications

The course of evolution has been affected by ice ages and other catastrophes on a global scale but much of the variation in life forms on Earth has resulted from more localized changes that have occurred throughout geological time. Isolation, besides the bringing together of previously isolated types, has had a major impact on the variation of species.

The separation of populations within a species prevents individuals breeding outside one particular population and hence prevents the mixing of genes and consequent variation. If such isolation is maintained, the species slowly diverges from its non-isolated members, as it becomes adapted to its own environment and the conditions peculiar to its habitat. Eventually it becomes physically different from other populations and may become incapable of breeding with them, even if contact is subsequently re-established. At this stage the isolated population has evolved into a new species.

Reproductive isolation refers to the inability of members of one population of a species to successfully interbreed with another. Such isolation ensures that the genes from one group will not be combined with those from another, thus each can accumulate its own genetic distinctions. There are several ways that groups can be reproductively isolated.

Geographical isolation

Periods of mountain formation, climatic changes, coastal erosion and the separation of land masses by continental drift all have a major impact on the development of new species. Obviously, if populations cannot reach each other, they cannot interbreed. There are many examples to illustrate this throughout the world from very diverse groups but some fresh water fish serve to illustrate the principle very well. In the USA large numbers of subspecies of rainbow trout are restricted to single lakes or rivers. They have all evolved from a single species which once was able to reach all of these now isolated regions when the geography of the area was different and the bodies of water were interconnected. The same is true of the char, a relative of the salmon, now isolated in a few British lakes as a result of changes produced by the last ice age glaciation.

In Australia, a group of birds called tree creepers are found around the coast. There are several species, each of which occupies a different area and is separated from its neighbours by dry or desert regions, which act as barriers and prevent cross-breeding. The Australian climate has changed several times in geological history and it is possible that at some time in the past one type of tree creeper was found throughout the continent. As the deserts developed and populations of this species became cut off from each other, they stopped interbreeding and eventually developed into distinct species.

The presence in Australia of marsupial mammals but no ancient placentals is also due to a physical change – that of the position of the continent itself. The marsupials originated in North America, from where they spread to South America and Europe. They became extinct in Europe and in North America because they were unable to compete with the placentals there. From South America, however, they moved across what was to become Antarctica and into the area that finally drifted apart to become the island continent of Australia. In the absence of competition from placentals they

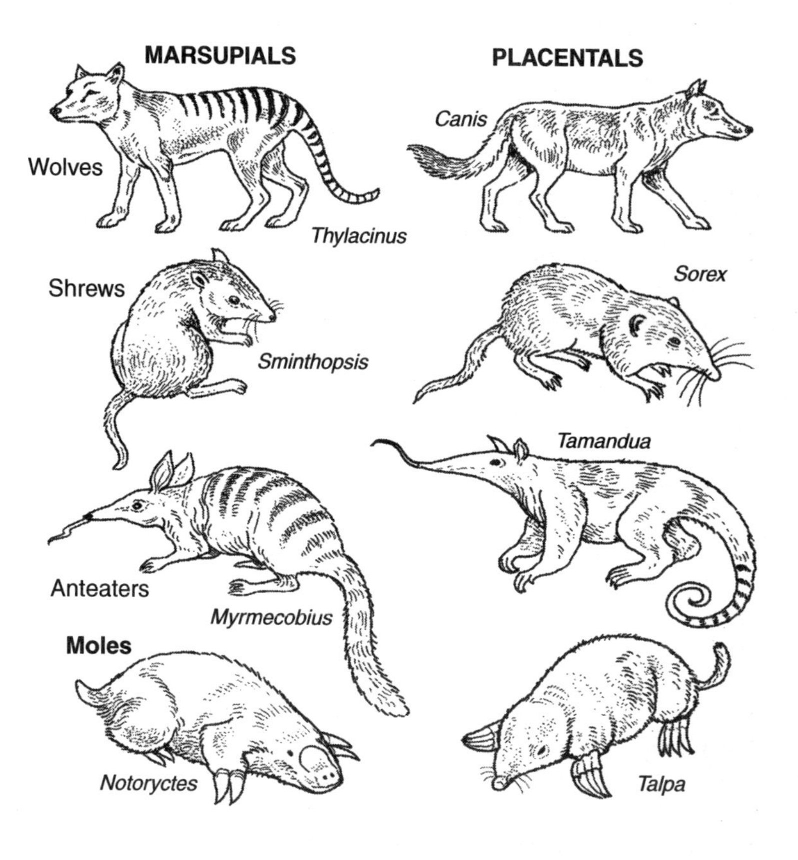

Figure 5.9 Adaptive radiation in placentals and mammals

were able to produce the great variety of forms that we see today, by divergent evolution, although a recent discovery of an ancient placental's tooth (see page 123) may disprove the theory of the *complete* absence of placentals.

Islands arise in a variety of ways but whether they are formed by coastal erosion, volcanic action or as coral formations they all have a profound influence on the evolution of organisms that live on them. Newcomers to recently formed islands often have a varied and previously unknown environment to colonize. Natural selection results in rapid change and diversity of the immigrants. On long-established islands natural selection has resulted in forms which are markedly different from the ancestral stock that arrived there. The birds occupy the niches that are available to birds everywhere but on an island they may all have descended from a single group. The Galapagos finches (see page 14) are classic examples. These small birds are found only on the Galapagos Islands and nearby Cocos Island, but are similar to birds on the south American mainland. They are classified into five genera, containing fourteen species. These differ from each other in feeding habit, which is reflected by the size and shape of their beaks. It is likely that a single flock of ancestral finches from the mainland somehow reached the Galapagos where, in the absence of competitors, they diversified rapidly.

It is very rare for large mammals to reach truly isolated islands, but if they do so it is likely that only the smallest endure, because creatures which make a heavy demand on their environment cannot survive in such restricted circumstances. A pygmy form of hippopotamus, for example, evolved on Mediterranean islands during the Pleistocene era, about 1 million years ago. On the other hand, dormice on these islands grew to a very large size, probably because a lack of predators made swiftness and agility unnecessary. Continental islands sometimes preserve forms that might have become extinct if they had met competition with later, more efficient, forms. The survival of the primitive reptile, the tuatara, which is the only survivor of the ancient group of reptiles called the *Rhynchocephalia*, has depended on its isolation on a few islands off New Zealand.

Behavioural isolation

The courtship behaviour of two groups may keep them from interbreeding even where their ranges overlap. For example, they may not recognize each other's mating signals or they may utilize different aspects of the environment in such different ways that their paths never cross. A variation in the breeding plumage of a bird, or a slight variation in its song might be enough to prevent mating. Even a change in biochemistry leading to a malfunction in the production of a chemical attractant such as a pheromone, can be equally important.

Mating isolation

Some species, especially insects have very complex 'lock and key' mating organs. They may not be able to copulate simply because their genitalia cannot fit together. Flowering plants, depending as they do on some outside agent for carrying pollen from one flower to another, have evolved along particular pathways which maximize their chances of pollination. This has often involved **co-evolution** between the plant and a particular insect which may be attracted to a flower and then carry out the pollination manoeuvres in a highly specialized way. Such specialization means that there is unlikely to be any chance of pollination occurring between different species.

Hybrid isolation

In cases where individuals of two species do interbreed the embryo may fail to develop because the mismatched chromosomes impede the normal processes of cell division which leads to growth. The **hybrids** which arise may be eliminated before they reach sexual maturity, and therefore make no contribution to future gene pools. Cross-fertilization is possible between closely related species such as goats and sheep but the hybrid embryos usually die very early unless they are produced as a result of *in-vitro* fertilization and then implanted into a genetically modified surrogate mother.

In some organisms hybrids may survive and even be fully fertile, but **reduced viability** means that they are unlikely to survive long enough to reach sexual maturity and to breed successfully. In waterfowl for example, fully fertile hybrids between mallards and pintails may be produced under laboratory conditions. The two species do not normally interbreed in nature, however, even though they may live and nest in the same very small area. The hybrids are less viable than their parents, possibly because they do not fit the ecological niche of either parental species, and so any hybrids which may be produced in the wild are unlikely to last long enough to reproduce.

Sterility of a hybrid will ensure that it cannot make a genetic contribution to future generations. The hybrid may be a physically strong individual but will still be unable to hand on any of its attributes. An example is the mule, which is a sterile hybrid between a male donkey and a female horse. Its sterility results from the inheritance of a single set of 32 chromosomes from its horse mother and a single set of 31 chromosomes from its donkey father. This total of 63 chromosomes makes it impossible for the production of sex cells by the hybrid because the full number of chromosomes must be halved during sex cell production. An uneven number makes this impossible.

So reproductive isolation then, is important in maintaining the integrity of a species. Let us now review some of the ways in which a species may arise.

A question of species

The apparently simple question 'What is a species?' has never been adequately answered. It is said that no two biologists will ever agree on an answer and therefore on a perfect description. If they appear to agree to disagree it usually means that they have exhausted their knowledge of exceptions to their definitions. So a lack of consensus is our starting point.

To the non-biologist the question of species seems easy because almost of us can tell the difference between a dog and a goldfish (the lead slips off a goldfish!). However, there are vast numbers of instances where organisms look so similar that they are very difficult to tell apart by external features and behavioural patterns – yet they never attempt to interbreed. They are quite different species and obviously have no problem in telling each other apart. For example, the golden-fronted woodpecker looks remarkably similar to the red-bellied woodpecker. The range of the former is from Honduras to central Texas, where it abruptly stops and the range of the latter begins. The red-bellied's range extends to the eastern coast, where it lives mainly in wooded habitats. The birds overlap in a narrow area of Texas but never interbreed to form hybrids. On the other hand, far more dissimilar species than these are known to interbreed. Examples are seen where coyotes or even wolves mate with domesticated dogs. So are the coyote and the domestic dog of different species? Since coyotes and dogs interbreed, even when differing greatly in appearance and behaviour, some scientists consider them to be the same species. Others argue that the *ability* to interbreed implies the *opportunity* to interbreed and that dogs and wolves or dogs and coyotes are sufficiently isolated that interbreeding between them is so rare that they can be considered different species. However, when they do interbreed the offspring are able to breed among themselves. This too, is a common criterion for animal species (but not necessarily for a plant species). Horses and donkeys can produce mules, but the mules are sterile, so the parents are considered to be different species.

These, then, are some of the problems when defining a species. The definition of the renowned zoologist **Ernst Mayer** will serve our purposes:

> A species is a group of actually or potentially interbreeding populations that is reproductively isolated from other such groups.

The ways in which specics arise are dependent on geography. Most species arise through **allopatric speciation** (*allo* = 'other'; *patric* = 'land'), which means the formation of new species after the geographic separation of once continuous populations. The process involves a population being somehow divided and each subgroup taking a different evolutionary route until they have diverged so much that interbreeding is no longer possible, even if they should rejoin.

Populations can be divided in two basic ways:

- One is a small group of individuals (or even a seed of a plant or a pregnant animal) being separated from the parent population and the descendants therefore becoming established in a new place. An example is Darwin's finches (see page 14) – they became **genetically isolated** not only from the parent population but also from each other as the various islands gave rise to their own distinct forms.
- Populations have also been divided by geological events, perhaps as some great barrier appeared separating a larger group into two smaller groups that then went into their own evolutionary ways. A example is seen in the Abert and Kaibab squirrels. These are two distinct species that live on opposite sides of the Grand Canyon. It is believed that they were once one population that was divided as the great chasm developed. The two populations followed their own paths of evolution and now are quite different and unable to interbreed.

The other type of speciation is called **sympatric** (*sym* = 'together'; *patric* = 'land'). It is less common than allopatric speciation and involves the formation of two species from one continuously interbreeding population. It is much more common in plants than in animals.

Among many flowering plants new species can arise by the interbreeding of existing species. This discovery surprised researchers since animal hybrids are usually sterile due to the inability of dissimilar chromosomes to carry out the normal process of sex cell formation. So why don't hybrid plants have the same problem with their chromosomes? The answer is primarily because plants with very different appearances may be very similar genetically. Where such genetically similar species overlap, there may be extensive hybrid populations. Surprisingly, such hybridization doesn't seem to result in the breakdown (through merging) of either parent species. This may be because hybrid populations find their own niche, interacting with the environment differently than the parent groups, thereby becoming truly a new and distinct species. Plants can also form new species in a quite dramatic way, where whole sets of chromosomes become doubled (or even doubled again). The condition is called **polyploidy** (*poly* = 'many'). Really it is a type of mutation (see page 161) where chromosomes double up by failing to separate properly during the normal process of sex cell formation. It can result in **tetraploid** (*tetra* = 'four') plants with four complete sets of chromosomes in each cell. Hybridization and polyploidy can create instant sympatric species of plants ready to be worked upon by the selective pressures of the environment. This versatility helps to explain how flowering plants arose so suddenly during evolution and how they then so quickly spread out over the land to become the amazingly diverse group that they are today.

Sympatric speciation in animals is much rarer than in plants but a few clear-cut examples have been observed. One of the best known cases involves flightless grasshoppers. The grasshoppers appear to be of one species across their range but chromosomal analysis reveals them to be of two different species, each occupying a particular part of the range. Apparently a random genetic change had occurred in the parent population that allowed the descendants of those carrying that change to better adapt to a part of the parent population's former range. The change seems to be great enough to preclude further mixing of the genes of the two populations, thereby giving rise to a new species in the midst of an existing one. This also serves as a good example of how modern developments in genetic profiling can be used in animals other than humans to show genetic similarities. An even more remarkable example of this technique being used to show sympatric speciation is the maggot fly. It has long been a pest in North America where it is a natural parasite of hawthorn trees. When apple trees were introduced to North America in the nineteenth century, the fly began to feed on these too. Now the fly has become divided into two groups of specialists, one preferring apple trees and the other remaining with its original diet, hawthorn. The two groups can still interbreed, but they differ in respect of several genes and in their maturation time.

In determining just how and when speciation has occurred it is important to realize that similarities may not be a reflection of evolutionary relationships. Just as different environments cause populations to diverge, a process called **divergent evolution**, sometimes similar environments can cause species to become more alike through a process called **convergent evolution**. Convergent evolution can cause a great deal of confusion for those trying to work out evolutionary histories if their studies rely on physical similarities. Far more valid conclusions may be gained via genetic profiling.

Great catastrophes

In the absence of environmental change it is possible to imagine a species evolving to the point where it is perfectly and completely adapted to its environment. It would remain unchanged, producing the occasional mutation which would die out quickly because, by definition, it could only be less than perfectly suited to its environment. Such situations do actually exist. For example, today's ocean-dwelling brachiopod *Lingula* has survived unchanged for 500 million years!

The ecology of any ecosystem is dynamic because one or more of its components (temperature, food supply, or level of predation) is always changing but, for a well adapted species, such regular cyclic fluctuations are already part of its evolutionary history and consequently its success. Such changes can be managed by the mechanisms that a species evolves to cope with them, such as reduced or increased population levels and overwintering.

When an environment changes gradually over many thousands of years, species can often adapt and evolve to keep pace with changing conditions. If a sudden extreme change occurs, however, organisms frequently cannot evolve fast enough and the species becomes extinct. When such a change is world-wide, mass extinctions may occur as **evolutionary catastrophes**.

The best documented of the physical changes is the occurrence of an ice age. From the time of the first invertebrates there have been at least four major ice ages, occurring at fairly regular intervals of about 150 million years and each lasting 10–50 million years. At such times the ice makes a number of advances and retreats, each advance being called a period of **glaciation**, and separated by warm **interglacial periods**.

About 18 000 years ago the spread of ice was at its maximum rate for the last period of glaciation. Glaciers advanced from the north and south poles over North America, Europe, and parts of Asia, covering at least three times as much of the Earth's land mass as they do today.

As ice advances on a global scale, the climate deteriorates, food becomes scarcer and environments become uninhabitable. Those animals that can migrate will move to other areas, which consequently puts a greater stress on food resources due to increasing competition. Ice ages are thought to have been accompanied by lowering sea levels. This factor would limit the diversity of habitats available to marine life, but possibly would also provide escape routes for land animals via land bridges. The plants, of course, could not escape. Consequently it is not surprising that ice ages are associated with mass extinctions, followed by rapid bursts of evolution as life colonizes and adapts to areas left by the retreating ice.

One such event was the Permo-Carboniferous Ice Age, which took place about 300 million years ago. The evidence of this is seen in rocks over huge areas of the southern hemisphere. Successive retreats of glaciers and a rise in the sea level drowned the swamps which gave rise to our coal measures, leaving behind layers of marine fossils in rocks which are otherwise full of terrestrial types. Many species became extinct during this time but others evolved rapidly in response to the climatic cooling, giving rise to the dominant vegetation of the southern hemisphere which flourished for more than 100 million years.

We live in an interglacial period of the latest ice age – the Cenozoic. Within the last 4 million years or so, the gradual expansion of ice cover and the periodic compression of climatic belts has been accompanied by the extinction and redistribution of animals and plants on a vast scale. As a result of such extinctions there are few large mammals left that were native to North America and Europe – Mastodons, mammoths, elephants, horses, giant beavers and sabre-toothed cats are just some of the types that now remain only as fossils.

A blast from the past

Perhaps the most well known puzzle related to geological catastrophes is the question of the sudden death of the dinosaurs. A global mass extinction at the end of the Cretaceous period, about 65 million years ago, affected a vast variety of species. However, it is mainly noted because of the disappearance of the dinosaurs.

Reasons for their extinction have fascinated biologists for many years and some spectacular theories have been put forward. For example, it has been suggested that:

- The newly developing small mammals ate the dinosaurs' eggs.
- Bombardment by cosmic rays made dinosaurs sterile.
- They could not adapt to changes in climate and vegetation.
- By their sheer size, the dinosaurs ate themselves out of existence or simply ran out of living space.
- Flowering plants were abundant at about this time, and many of the new plants contained chemicals which may have been poisonous to herbivorous dinosaurs, which in turn were the chief food of the carnivorous types.
- Newly evolved insects might also have competed with herbivorous type for food.

Of all the theories put forward, the least unlikely concerns climatic change. At the end of the Cretaceous there was a change from a relatively stable world-wide climate to more local extremes with regions where the only surviving species were those that were able to adapt to new conditions quickly enough.

The most widely held current theory postulates an enormous impact by a meteorite. In fact it is now so well established that some scientists consider it to be irrefutable. In 1980, the discovery of the iridium layer at the Cretaceous–Tertiary boundary by **Luis** and **Walter Alvarez** provided evidence of such a massive impact. Iridium and osmium are very rare elements on Earth and are said to have come from extraterrestrial sources like meteorites and asteroids. 65 million years ago a meteorite of gigantic proportions crashed into the Earth and threw up an enormous mass of dust, which must have been impenetrable to sunlight for such a long time that food plants died. Without the food plants as producers, herbivores died out and the knock-on effect on food chains led to mass extinctions. In the mid 1960s, a British schoolteacher, **Joe Enever**, published a detailed description of the effect of a meteorite impact. The inspiration for Enever's interest was remnant scars and impact structures that the Earth bears from the remote past.

Enever's ideas made little impression at the time, but since the 1980s much research has been carried out to investigate the catastrophic potential of meteorite impacts. The impetus for such work was largely due to the identification of the 180-km wide Chicxulub crater in the Yucatán Peninsula in Mexico, which is now firmly established as the collision site which blew away the dinosaurs. Such a crater could be explained by an impact of an asteroid about 10 km across. In 1996 **Peter Schultz** of Brown University in Providence, Rhode Island and **Steven D'Holt** of the University of Rhode Island, provided evidence to show that the fateful asteroid hit the peninsula from an angle of 20–30° above the southern horizon, showering molten debris towards the north-west. The greatest effect was seen in western North America but its legacy was world-wide (Figure 5.10). The angle of the impact was crucial because it affected the distribution of energy release. A direct, full-frontal impact would concentrate energy downward into the bed rock but an angled one would scatter debris along the missile's path. Schultz suggested that such a collision would have generated a massive cloud of vaporized and molten rock, saturating the atmosphere over the west of North America and instantly roasting any living thing in its path. Fragments ejected by the impact would have re-entered the atmosphere like smaller meteorites, with further devastating consequences. According to the calculations by **Jay Melosh** and colleagues at the University of Arizona, the blast could have stripped away some of the Earth's atmosphere into space, while the energy of the particles re-entering would have heated up what was left by 10 kW/m^2 of the Earth's surface for several hours after the impact. This level of heating helps to explain one of the most striking features of the geological boundary at the end of the Cretaceous – a soot layer produced by a global firestorm of mind boggling extent.

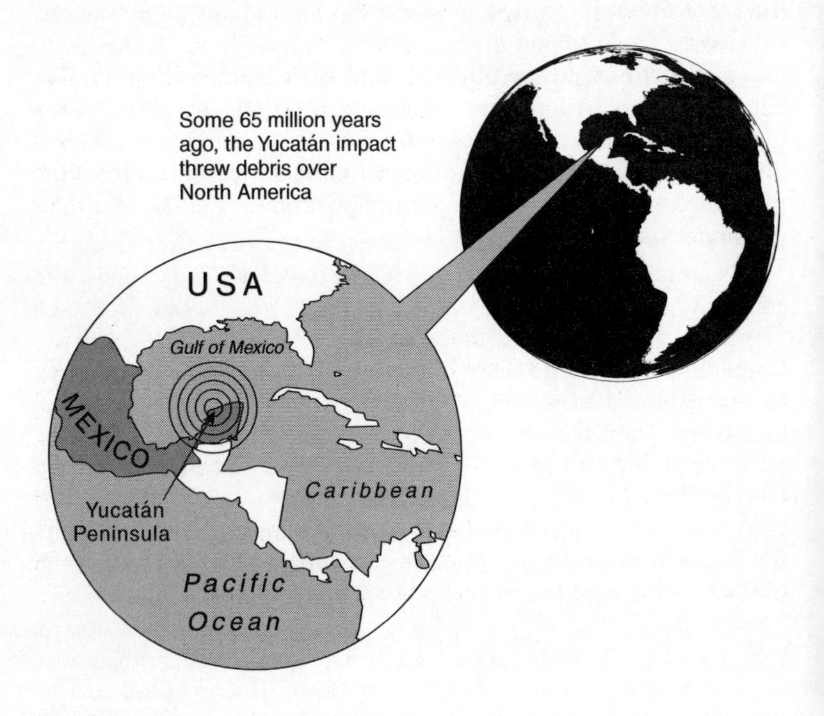

Figure 5.10 The one that blew away the dinosaurs

Computer-generated simulations have estimated that winds of at least 500 kph would have been developed by the impact, resulting in the destruction of vast areas of trees. The heating effect suggested by Melosh's team would have ignited all combustible material producing 70 billion tonnes of soot, equivalent to the burning of 25% of the total organic matter estimated to have been present on Earth at the time. The direct effects of the fires alone would have included a period of darkness and cold caused by the smoke. Nitrous oxide and mutagens caused by the fires would have had disastrous consequences for living organisms, as would a build-up of concentration of carbon dioxide in the atmosphere to at least three times its present concentration. Strong winds and enormous destructive waves, tsunamis, would have raged for hours and the fires for months. The darkness and cold would have lasted even longer. Poisons and mutagens would have remained active for years, as would the effects of acid rain and the destruction of the ozone layer. Also the frequency of volcanic activity would have increased, triggered by the impact. It is not really surprising, therefore, that dinosaurs, along with most other species, met their end under such conditions!

Conclusion

Evolution, as we understand it today, is simply a function of basic mathematics: those genes that promote successful reproduction will increase in frequency. Life on Earth is subject to countless pressures as it continues striving for its very existence. It must constantly react to the nature of its situation (or its predicament) and it must change. The world is a variable and changeable place, and different life forms have evolved that are uniquely able to utilize one aspect of the Earth or another in countless ways. In a nutshell:

> life must change in order to take advantage of its part of the world.

Biologists are no longer looking for proof of evolution, only for advanced knowledge of its complex process or for the detailed history of certain animal and plant lines. Yet many non-biologists allow themselves to wonder audibly whether Darwinism is still believed, or if the 'theory' of evolution is all it was thought to be. The answer is an emphatic 'Yes'. The masses of fossilized bones and teeth, and the long shelves of biological studies, are not the only cornerstones of our understanding. The two principal foundations are the contributions of Darwin and Mendel, and they consist respectively of a largely irresistible force and a moderately movable object. The object, which Mendel did so much to explain, is heredity. The force – Darwin's monument – is natural selection. The combination is evolution.

The geological time scale

Millions of years ago

Eon

- 0– Phanerozoic
- 540– Proterozoic
- 2500– Archean
- 4000– Hadean
- 4550–
- Precambrian (540–4550)

Events on the Eon scale:

- First hard-shelled animals
- First complex animals
- Eukaryotic cells
- First free oxygen
- Stromatolites
- Earliest life
- Oldest rocks (4000)
- Oldest Zircon crystals (4200)
- Formation of Earth (4550)
- Formation of the Sun

Era

- Cenozoic Tertiary
- Mesozoic
- Palaeozoic

Period

Period	Millions of years ago
Holocene	0.01
Pleistocene	2
Pliocene	5
Miocene	24
Oligocene	37
Eocene	58
Paleocene	65
Cretaceous	144
Jurassic	202
Triassic	245
Permian	286
Carboniferous	360
Devonian	410
Silurian	433
Ordovician	505
Cambrian	540

Extinctions and victims

- Large mammals
- Mammals
- Mammals, shelly marine animals
- Dinosaurs, marine reptiles, pterosaurs, shelly marine animals, ammonoids, reef-building clams
- Shelly marine animals
- Shelly marine animals, dinosaurs, ammonoids
- Shelly marine animals, reef builders, mammal-like reptiles, ammonoids, conodonts, snails
- Shelly marine animals, reef builders, mammal-like reptiles, ammonoids, last trilobites
- Shelly marine animals, reef builders, trilobites, primitive fish, ammonoids
- Shelly marine animals, reef builders, trilobites, nautiloids
- Trilobites
- Small shelly animals, reef builders, trilobites

Evolutionary events

- Early humans
- Whales, bats, modern mammals
- Flowering plants
- Earliest birds
- First dinosaurs and mammals
- Reptiles
- Amphibians on land
- Land plants
- First (jawless) fish
- First diverse animals

GLOSSARY

Adaptation Any structure, physiological or behavioural feature which enables an organism to cope with a change in its environment.

Adenosine triphosphate (ATP) An organic compound capable of storing energy within the cell in the form of high energy phosphate bonds.

Aerobic In the presence of oxygen.

Agnathostome The class containing the earliest and most primitive vertebrates, characterized by the absence of jaws (*a* = 'without'; *gnath* = 'jaw'). Living types include the lampreys and hagfish, the descendants of a primitive group of fish-like creatures, the ostracoderms.

Algae A class of plants including the seaweeds and the fresh water weeds made of filaments.

Alternation of generations This term is applied to a peculiar mode of reproduction which prevails among many of the lower plants and animals. It is the alternation of a generation which reproduces sexually with a generation which reproduces asexually.

Amino acid A subunit of a protein.

Ammonites A group of fossils with spiral, chambered shells, allied to the existing pearly nautilus, but having the partitions between the chambers waved in complicated patterns at their junction with the outer wall of the shell.

Amphibians A class of cold-blooded vertebrates with soft slimy skins. They nearly all lay eggs in water and have aquatic larvae which change into adults which generally live part of their lives on land.

Amphioxus A marine animal which inhabits shallow water and externally resembles a small fish without fins. It belongs to a group which may be ancestral to the vertebrates, having a stiffening rod called a notochord, rather than a vertebral column. It also lacks a skull.

Analogous structures Those structures found on or in organisms which have similar functions but not similar ancestries, for example the wings of insects and those of birds.

Anapsid A type of reptile's skull in which there are two enlarged openings, but a solid sheet of bone on top and at the sides.

Annelids A class of worms which includes earthworms, leeches and marine worms which are segmented with a ringed appearance. (From *annulus* = 'a ring').

Antennae Jointed organs appended to the head in insects, crustacea, centipedes and millipedes.

Anthers The summits of the stamens of flowers, in which pollen is produced.

Arthropoda A large division of the animal kingdom, characterized generally by having the surface of the body divided into segments, some of which have jointed legs. It includes insects, crustaceans, centipedes and millipedes. (From *arthro* = 'jointed'; *podum* = 'foot').

Asexual reproduction Reproduction without involving the fusion of sex cells.

ATP See adenosine triphosphate.

Australopithecus A genus of 'ape-men' or hominids related to *Homo sapiens*.

Bipedal Walking on two limbs rather than on four.

Bird-hipped The type of hip found in some dinosaurs with a superficially bird-like arrangement of bones.

Bivalve Mollusc with two shell valves hinged together – examples are cockles, mussels, clams.

Brachiopoda A phylum of marine animals with bivalve shells, attached to a submarine object by a stalk. They have cilia which are used in feeding.

Cell The units which make up living things, separated from each other by membranes.

Cephalopoda The class of molluscs to which the octopus, squid, nautilus and ammonites belong.

Chitin A nitrogen–carbohydrate derivative which forms part of the outside skeleton of invertebrates, especially arthropods.

Chlorophyll The green pigment in plant cells that is essential for photosynthesis.

Chloroplast A membrane-bound structure in plant cells which stores chlorophyll.

Chromosomes Strands of genes which are made of DNA. They are found in the nuclei of cells.

Cilia Minute, movable hair-like structures protruding from the surface of some animal cells. They help in movement of single-celled animals or they help the movement of fluids in multicelled types.

Coacervates Structures postulated as forming during the production of the first living matter from non-living matter.

Cold blooded An animal unable to maintain a constant body temperature.

Continental drift The slow movement of the continents across the face of the globe, so that their relative positions change over long periods of geological time.

Continental shelf The offshore edge of a continent, gently sloping down to about 200 m.

Convergent evolution If two forms are descended from very different ancestors but show a superficial similarity to each other through adaptation to similar niches, they are said to show convergent evolution. Examples are whales and fishes.

Crossing over The exchange of parts of chromosomes during the type of cell division which produces sex cells.

Crustaceans A group of jointed-legged invertebrates with an outside skeleton which is calcified. They have two pairs of antennae and breathe using gills.

Cycads Type of superficially palm-like shrubs and trees, abundant in the Mesozoic, and still common today in the southern hemisphere.

Deoxyribonucleic acid (DNA) A complex molecule in the shape of a double helix (spiral). It is capable of replicating itself and is the basis of the structure of a gene and the genetic code.

Diapsid A type of reptile skull in which there are two large 'windows' in the superficial sheet of bone. Lizards and dinosaurs have this type of skull.

Diffusion The net flow of molecules from where they are in high concentration to where they are in low concentration down a gradient.

Dinosaur Reptiles of the extinct orders Saurischia and Ornithischia, abundant during the Mesozoic.

DNA See deoxyribonucleic acid.

Dominant character One of a pair of contrasted characters which shows itself in the heterozygous (hybrid) condition.

Dorsal Of, or belonging to, the back.

Ecology The study of the interrelationships between living organisms and their environment.

Ecosystem A unit made up of living and non-living components of a particular area that interact and exchange materials with each other.

Embryo The young organism undergoing the first stages of development in the egg, seed or womb.

Embryology The study of the development of organisms.

Environment The surroundings and conditions within which an organism lives. It includes the influence of physical conditions such as climate and also of other organisms living there.

Enzymes Biological catalysts or proteins which speed up chemical reactions taking place in organisms.

Era A major unit of geological time containing several periods.

Eukaryote An organism which has cells in which the chromosomes are in the nucleus, surrounded by a nuclear membrane. They include all organisms except bacteria and blue–green algae which are known as prokaryotes.

Exoskeleton A skeleton which is on the outside of the body, as are the hard parts of insects and crabs.

Extinction The total disappearance of a species, whether through dying out of all individuals or by evolution to another species.

Fauna The totality of animals naturally inhabiting a certain region or which have lived during a geological period.

Flora The totality of plants naturally growing in a certain region or which have lived during a geological period.

Fossil The remains of, or impressions left in rocks by, long-dead animals and plants.

Gene The smallest indivisible unit of heredity located on a chromosome and made from DNA. Genes may exist in a number of forms called alleles.

Genetics The study of heredity and variation.

Ginkgo The only remaining genus of 'maidenhair' tree. A diverse group of trees in the Mesozoic but now is a 'living fossil'.

Glaciation The formation of glaciers. Each ice age is called a glaciation.

Habitat The locality in which a plant or animal naturally lives.

Heterozygous A heterozygous animal or plant is an individual which has received unlike genes from both its parents.

Hominid Member of the family of humankind, Hominidae.

Homozygous An individual which has received identical genes from both its parents (pure bred).

Hybrid The offspring of the crossing of two distinct species.

Industrial melanism The occurrence, in areas of air pollution of black varieties of species which are normally a lighter colour.

Invertebrate Any animal without a backbone.

Keratin A tough fibrous protein that makes up hair, nails, horns and hoofs.

Lancelet See Amphioxus.

Land bridge An area of land, temporary on the geological time scale, that joins two larger, more permanent, land areas and provides a corridor for the spread of animals and plants. Examples have occurred between Britain and the rest of Europe and between North America and Russia.

Larva The first condition of an animal which hatches from an egg if it is to undergo a change in form to become an adult – for example, a maggot, caterpillar, tadpole.

Lipid Any fats or waxes found in living tissue.

Lycopoda A large group of plants, including trees in the carboniferous, and today represented by club mosses.

Mammalia Warm-blooded vertebrates with hair on their bodies at some stage in their development and which suckle their young by producing milk.

Marsupials Mammals which retain their young in pouches while they are suckling.

Meiosis The form of cell division in which there is a reduction by half of the normal number of chromosomes of a species. It results in the formation of sex cells.

Melanism The development of an unusually large amount of brown pigment, melanin in the skin. The opposite of albinism.

Membrane A thin layer enclosing cellular contents or a tissue covering an organ.

Metabolism The sum of all the chemical reactions going on in an organism.

Missing link An animal or a plant that forms an evolutionary link between one major group and another. It can be a misnomer if a known organism is called one.

Mitosis The normal form of cell division which gives rise to identical cells and is responsible for growth.

Molluscs Soft-bodied invertebrates, unsegmented, with most body organs in a dorsal position and usually covered by one or more shells. Clams, snails, ammonites are all molluscs.

Monotremes Egg-laying mammals.

Multicellular Made of many cells.

Mutation A spontaneous change in the structure of a gene or a chromosome, resulting from an inaccurate copying of DNA or by damage due to mutagens like ionizing radiations or certain chemicals.

Natural selection An evolutionary mechanism by which the total environment selects those forms that are best suited to survive and to breed.

Niche An organism's source of food and shelter, its interaction with other organisms, and all other factors that define its position in the ecosystem.

Nucleus An area within a complete eukaryotic cell, enclosed by a membrane, and containing the chromosomes.

Osmosis The net flow of water molecules through a selectively permeable membrane from a region of high water concentration to a region of low water concentration until dynamic equilibrium is obtained when the passage of the water is equal in both directions across the membrane.

Palaeontology The study of fossils.

Pangaea The Permian–Triassic supercontinent.

Parallel evolution If two forms are derived from a common ancestor but have independently evolved very similar adaptations to similar ecological niches, they are said to show parallel evolution.

Period A division of geological time often marked by a particular abundance of one group of plant or animal. Periods are joined together as eras.

Photosynthesis The process by which plants convert the energy from sunlight into chemical energy in food by using carbon dioxide, water and chlorophyll.

Placental mammals Mammals which have a placenta to nourish their developing embryos. The placenta is an organ which establishes close contact between the blood supplies of the mother and the embryo for exchange of materials by diffusion.

Pollen Particles produced by the anthers of flowering plants which contain the male sex cell.

Population A discrete group of interbreeding animals or plants within an isolating boundary.

Primitive Originating very early, describing the first plants or animals. Also used comparatively of forms of earlier origin than others and which lack advanced features.

Prokaryote Organisms lacking a well defined nucleus in their cells. Examples are the bacteria and blue–green algae.

Protein A class of molecules made up of units called amino acids. Some make up structures of organisms, others are used as chemicals to control reactions, e.g. enzymes and some hormones.

Protozoa Single-celled animals.

Recessive character One of a pair of contrasted characters which shows itself only in the homozygous condition.

Ribonucleic acid (RNA) A chemical found in all living cells which occurs in three separate forms and which controls the synthesis of proteins in cells.

Sediment A loose material weathered from rocks and soil and transported by water or wind until it is deposited. It may consolidate into sedimentary rocks.

Selective pressure The effects of natural selection when they are tending in one particular direction to affect survival of a species.

Species A group of individuals that can interbreed and produce fertile offspring.

Stamens The male sex organs of flowering plants.

Supercontinent A giant land mass formed by the coming together of continents through continental drift.

Swim bladder A buoyancy organ found in fish with bony

skeletons, derived from a pouched outgrowth from the gut. In many primitive fish it may have functioned as a lung.

Synapsid A type of reptile skull which has a single 'window' low in the roofing bone at each side. The reptile ancestors of mammals had this type of skull.

Taxonomy The science of classifying organisms into groups.

Trilobites A peculiar extinct group of marine animals resembling crustaceans in external form. They are found as fossils most abundantly in Silurian rocks.

Vertebrate Animals with backbones (vertebrae).

Vestigial A part of an organism which was functional in its ancestor but is reduced or functionless in present day types.

Warm blooded Any animal that can maintain a constant body temperature.

FAMOUS NAMES

The following is a list of names of some famous original thinkers and contributors to studies of evolution and the origin of life.

Aristotle (384–322 BC) Among the greatest of all biologists. He arranged and classified existing knowledge and added to it by describing the habits of many marine animals.

Robert Broom (1866–1951) Discoverer of fossils of several subspecies of *Australopithecus* in the 1940s.

Georges Cuvier (1769–1832) A supporter of special creation and who explained the presence of fossils by new life being created and then buried by successive cataclysms.

Raymond Arthur Dart (main contributions in the 1920s and 1930s) Discoverer of the fossil remains of the 'Ape of the south', *Australopithecus* in 1925.

Charles Darwin (1809–1882) Advanced the theory of natural selection as the basis of evolution. The most famous evolutionary biologist of all time.

Erasmus Darwin (1731–1802) Suggested that species are not immutable and possibly acted as a catalyst for his grandson, Charles Darwin's, thinking.

George-Louis Leclerc de Buffon (1707–1788) A believer in special creation but suggested that monkeys developed from humans as a result of imperfections in the Creator's expression of the ideal.

Eugene Dubois (1858–1940) The first to discover the fossil remains of Java Man, *Homo erectus*.

Gustav Heinrich Ralph von Koenigswald (main contributions in 1930s and 1940s) Discovered fossilized teeth of the 'giant ape' *Gigantopithecus* in 1935 and contributed to the discoveries of fossils of *Homo erectus*.

Jean Baptiste Lamarck (1744–1829) Attempted to explain evolution by suggesting that there is inheritance of characteristics acquired during a lifetime by use and disuse.

Louis Seymore Bazett Leakey (main contributions in the 1940s and 1960s) Discoverer of the fossil remains of the 'missing link' between apes and humans in the form of Proconsul in 1959.

Antony van Leeuwenhoek (1632–1723) The first person to examine and describe microbes accurately with the use of a microscope.

Carolus Linnaeus, born Carl von Linne (1707–1778) Invented the binomial system of naming plants and animals which is still used in its established form when classifying the living world.

Charles Lyell (1797–1875) In his *Principles of Geology* was the first to recognize the true age of the Earth. Influenced Charles Darwin's theory of natural selection.

Thomas Malthus (1766–1834) Produced the first published warning of the dangers of over-population, sowing the seeds for Charles Darwin's idea of the survival of the fittest to breed.

Gregor Mendel, born Johann Mendel (1822–1884) Established the fundamental laws of inheritance which are used to explain the genetic basis of evolution through variation.

Stanley Lloyd Miller (main contributions began in 1950s) The first scientist to simulate atmospheric conditions of early Earth and produce amino acids from simpler chemicals.

Alexander Ivanovich Oparin (main contributions late 1930s to 1960s) The first scientist to study the problem of the origin of life in great detail in his book *The Origin of Life*, published in 1936 and translated into English in 1938.

Francesco Redi (1626–1697) One of the first investigative biologists who discredited the idea of spontaneous generation by demonstrating that maggots developed from the eggs of flies, rather than from rotten meat.

Lazaro Spallanzani (1729–1799) The first person to demonstrate experimentally that microbes do not originate spontaneously.

Alfred Russel Wallace (1823–1913) Arrived at many of Charles Darwin's conclusions independently. Jointly, with Darwin, presented research papers on natural selection to the Linnean Society, London, in 1858.

Alfred Wegener (1880–1930) Put forward the idea of continental drift in his book *The Origin of Continents and Oceans* in 1915.

FURTHER READING

Brookfield, A.P. *Modern Aspects of Evolution* (Hutchinson, 1986).
Darwin, Charles. *The Origin of Species* (John Murray, 1859).
Darwin, Charles. *The Descent of Man* (John Murray, 1871).
Darwin, Francis. *The Life of Charles Darwin* (John Murray, 1902).
Dawkins, R. *The Selfish Gene* (Oxford University Press, 1976).
Dawkins,R. *The Blind Watchmaker* (Longman Scientific and Technical, 1986).
Dowdswell, W.H. *Evolution : a modern synthesis* (Heinemann Educational, 1984).
Fortey, Richard. *Life: An Unauthorised Biography* (HarperCollins, 1997).
Gould, Stephen Jay. *Wonderful Life: The Burgess Shale and the Nature of History* (Hutchinson Radius, 1989).
Jenkins, Morton. *Inheritance and Selection* (Simon and Schuster, 1992).
Jenkins, Morton. *Teach Yourself Genetics* (Hodder & Stoughton, 1998).
Leakey, Richard. *The Origin of Humankind* (Weidenfeld & Nicolson,1994).
Leakey, Richard and Lewin, Roger. *The Sixth Extinction: Biodiversity and its survival* (Weidenfied & Nicolson, 1996).
Lewin, Roger, *The Origin of Modern Humans* (W.H. Freeman, 1993).
Miller, S.L. and Orgel, L.E. *The Origin of Life on Earth* (Prentice-Hall, 1974)
Rose, Steven. *The Chemistry of Life* (Penguin, 1991).
Rose, Steven. *Lifelines* (Penguin, 1997).
Smith, John Maynard. *Did Darwin Get It Right?* (Penguin, 1993).
Walker, Alan and Shipman, Pat. *The Wisdom of Bones: In Search of Human Origins* (Weidenfeld & Nicolson, 1996).

INDEX